Practical Lambing and Lamb Care
A Veterinary Guide
Second Edition

Practical Lambing and Lamb Care
A Veterinary Guide
Second Edition

Andrew Eales BSc BVSc MSc PhD DSHP MRCVS

and

John Small HNC

with drawings by

David Pollock SDA

Longman
Scientific &
Technical

Longman Scientific & Technical
Longman Group Limited
Longman House, Burnt Mill, Harlow
Essex CM20 2JE, England
and Associated Companies throughout the world

First published 1986

Second edition 1995

British Library Cataloguing in Publication Data
A catalogue entry for this title is available from the British Library.

ISBN 0-582-21004-6

Set in 10/12 pt Sabon
Produced by Longman Singapore Publishers Pte Ltd

Contents

Preface to the first edition

Each year millions of newborn lambs and tens of thousands of ewes die at lambing time. This level of loss is unacceptable and management from before tupping to lambing itself must be aimed at reducing it to an absolute minimum. This book is concerned with one small but important part of this management – veterinary care of lambing.

We have addressed ourselves to those who care for lambing ewes and newborn lambs, principally shepherds, and to others closely concerned with lambing management: veterinary surgeons, agricultural and veterinary students, and agricultural and veterinary teachers.

To some extent this is a 'how you do it' text but we have tried to introduce an understanding of the various problems and techniques discussed. Thus in the first chapter we have concentrated on the factors which make newborn lambs apparently so prone to problems in the first few days of life. The prevention of problems in newborn lambs depends on an understanding of these factors.

We have assumed throughout that the shepherd has a sympathetic relationship with his veterinary surgeon. It is, of course, totally impractical for the veterinary surgeon to attend each difficult lambing, each sick ewe and each sick lamb. The cost would be prohibitive, treatment in many cases would inevitably be given too late and there are not enough vets to go around anyway. This means that the veterinary surgeon must assume the role of consultant. Before lambing the problems likely to be encountered should be discussed, treatments defined and techniques learned. The role of the veterinary surgeon does not cease here. New problems will arise and the treatment of routine problems should be monitored. An occasional visit during lambing is most helpful. On-going problems can be reviewed and mental notes made for improvements in the future.

This text may be used solely as a 'first-aid' reference for when things go wrong but to get maximum benefit we would suggest that it is read before lambing. This should ensure that the necessary equipment is to hand and that valuable time is not wasted looking for the appropriate section by torchlight at three o'clock in the morning!

In writing this book we have drawn freely on the work of our colleagues at the Moredun Research Institute and many friends and colleagues from the Agricultural Development and Advisory Service, the Hill Farming Research Organisation, the Meat and Livestock Commission, the Rowett Research Institute, and the Scottish Agricultural Colleges. We hope our text does full justice to their work.

We thank Dr W. B. Martin DVSM PhD MRCVS FRSE, Director of the Moredun Research Institute, for encouragement throughout this project.

We are indebted to B. J. Easter C&G Adv and A. Inglis C&G Adv, whose photographic skills are to be found throughout the text.

We thank the undermentioned who reviewed the original draft and made many helpful and constructive suggestions: R. M. Barlow DSc DVM&S MRCVS; W. S. Dingwall BSc PhD; J. FitzSimons NDA; J. S. Gilmour BVM&S FRCVS; Lorna A. Hay BVMS MRCVS; D. C. Henderson BVM&S MRCVS and G. E. Jones BVSc DTVM PhD MRCVS.

The drafts were most ably typed by Mrs R. Cannel, Miss J. Goodier and Mrs K. Mark.

Finally we received unfailing support from our families throughout the composition of this work and we offer them our sincere and humble thanks.

ANDREW EALES
JOHN SMALL
Moredun Research Institute, Edinburgh
December 1984

Preface to the second edition

The reception of the first edition of this book has been most gratifying and has encouraged us to proceed to this second edition.

In this new edition we have incorporated recent developments, of which there have been many, and also extended the scope of the text to include infectious abortion in ewes and problems in lambs to weaning. We hope our readers will find these additions helpful.

We have also added a new chapter on 'Welfare at Lambing'. Lambing is a busy time for all shepherds and sheep farmers, but it is also the time when most problems arise. Understaffing and tiredness can easily lead to problems missed and unnecessary suffering. A little forethought will prevent most of these problems.

We thank B. J. Easter C&G Adv. for his skilful help with photography.

We are grateful to our friends at Long Yester and Sourhope for help with photographic facilities.

We are indebted to Dr David Buxton BVM&S PhD FRCPath MRCVS for reviewing our new material, and making many helpful and constructive comments.

The drafts were most ably typed by Mrs J. Whitehouse and Mrs S. McGill.

We thank Christine and Vasanthi for carefully checking the drafts, and for their forbearance during our many 'editorial meetings'.

Finally we are grateful to our publisher, Longman Higher Education, for their support, understanding and co-operation in this venture.

ANDREW EALES, JOHN SMALL
July 1993
Edinburgh

Acknowledgements

We are indebted to the following for permission to reproduce
copyright material:

Dr R. M. Barlow for Fig. 3.17; Dr R. M. Barlow, W. B. Martin and
Blackwell Scientific Publications Ltd for our Fig. 3.1 from *Diseases of
Sheep* (ed. W. B. Martin); M. J. Clarkson for Plate 10; Council of the
Scottish Agricultural Colleges for our Figs 4.5 and 7.24 from
Management at Lambing (Publication No. 22, 1983), our Figs 7.15 and
7.18 and extracts from pp. 3–6 of *Hypothermia in the Young Lamb*
(Technical Note No. 60, 1983); Dalton Supplies Ltd, Nettlebed, for
Fig. 7.22; the late J. S. Gilmour for Figs 1.7 and 1.8; Hill Farming
Research Organisation and the *Veterinary Record* for our Fig. 6.2
from *In Practice* 6(3) 1984, p.92; Dr. R. H. F. Hunter, Longman Group
and Academic Press for our Fig. 1.5 from *Reproduction of Farm
Animals* (Longman 1982) and *Physiology & Technology of
Reproduction in Female Domestic Animals* (Academic Press 1980); A.
Inglis for Plate 6; K. A. Linklater for Plate 4; Longman Group Ltd for
our Fig. 6.1 from *Sheep Production Science into Practice* by A. W.
Speedy; D. J. Mellor for Plate 1; Ministry of Agriculture, Fisheries and
Food for extracts from the *Codes of Recommendation for the Welfare
of Livestock: Sheep* (©Crown Copyright 1990); Modulamb Ltd for
Fig. 7.23; Meat and Livestock Commission for an extract from *Feeding
the Ewe* (1983).

To the memory of our late friend and colleague
John S Gilmour BVM & S MRCPath FRCVS

Chapter 1
The newborn lamb

What makes the newborn lamb prone to so many problems? This is a complex subject and researchers are still looking for many of the answers. Research has, however, revealed much that is useful to those of us who work at the 'sharp end'. In the next few pages we have summarised this knowledge. First we shall look at what we might call the 'perfect' lamb, such as a good single out of a mature ewe in good body condition, and see how this lamb is disadvantaged when compared with adult sheep. Then we will examine the various factors which can cause newborn lambs to be less than 'perfect' and more prone to problems in the first few days of life.

Table 1.1 The relative importance of the different causes of lamb death

Foetal stillbirth (death before lambing)	10–20%
Parturient stillbirth (death during lambing)	10–20%
Hypothermia due to exposure	15–25%
Hypothermia due to starvation	20–30%
Infectious disease	10–15%
Congenital abnormalities	c. 5%
Other causes	c. 5%

Problems facing the 'perfect' lamb

Most problems in newborn lambs are associated with either nutrition, temperature regulation or infectious disease and it is useful to consider the differences between the adult and the newborn under these three headings.

Nutrition

When considering nutrition in the newborn lamb we are mainly concerned with energy. Protein and other nutrients are of course essential for growth but we are most interested in survival for the first few days of life and it is a shortage of energy which is most likely to reduce viability. When compared with the adult sheep the newborn lamb has three problems.

1. The lamb has lower energy reserves in its body in the form of stored fat and carbohydrate. Total energy reserves in the newborn lamb only account for about 3 per cent of body weight – the corresponding figure in the adult sheep is 10–15 per cent.
2. Whereas the adult sheep is 'self-feeding' providing fodder is available, the newborn lamb is totally dependent on its mother for its food supply.
3. The newborn lamb needs more energy than the adult sheep on a body weight basis. This statement requires a little explanation. The most important use of energy in any mammal in a cold climate is the maintenance of body temperature – 'keeping warm'. An animal loses heat mainly through the skin, and if the body temperature is to be maintained heat must be produced to equal this heat loss. The crucial point is that the newborn lamb has, in proportion to its body weight, considerably more skin than the adult sheep and thus proportionately it loses more heat. To give an example: a 4 kg lamb has proportionately three times more skin than a 60 kg adult sheep. This means that in proportional terms it will lose three times as much heat and to maintain its body temperature it will have to produce three times as much heat. To do this it needs three times as much energy, i.e. food.

In summary, the lamb has limited energy reserves stored in its body, is totally dependent on its mother for its energy supply and, in proportion to its body weight, needs considerably more energy than the adult sheep. It is not surprising that starvation is a major killer of newborn lambs.

Temperature regulation

We have seen already that a lamb must produce as much heat as it loses if it is to maintain its body temperature. If a lamb either loses too much heat or cannot produce enough heat its body temperature will fall – hypothermia – and it will die.

Let us first consider the problem from the heat-loss point of view. We already know that the newborn lamb has proportionately more skin than the adult sheep and so has an inherent higher rate of heat loss. There are two further factors which increase the rate of heat loss from newborn lambs:

1. The birth coat has a low insulation value when compared with the full fleece of the adult sheep.
2. The newborn lamb is wet when it is born. This not only reduces the insulation value of the fleece but also leads to a high rate of heat loss caused by the evaporation of water from the coat, especially in windy conditions. Anyone who has stepped out of a piping hot bath into a cold draughty bathroom will appreciate this problem.

The ewe plays a very important part in reducing the rate of heat loss from the newborn lamb. The faster she licks her lamb dry, the lower is the rate of heat loss and the risk of hypothermia. Shelter also reduces the risk of hypothermia and stone dykes or walls of straw bales greatly moderate the effects of a strong wind. Housing, of course, is the ultimate form of shelter.

These three factors – a large area of skin through which to lose heat, a birth coat of poor insulation value, and being born wet – all add together to make the newborn lamb highly susceptible to hypothermia due to exposure in the first five hours of life. Hypothermia during this period probably accounts for one-quarter of all lamb losses.

The 'perfect' newborn lamb is an excellent generator of heat. A 6 kg lamb can produce as much heat as a 100 watt light bulb! BUT a high rate of heat production can only be maintained if energy is available. If a lamb starves, its body energy reserves quickly become exhausted and heat production practically stops. Hypothermia, in this case caused by starvation, is the inevitable result, even in a warm environment such as a sheep house. Lambs can die from hypothermia due to starvation before they are twelve hours old. This problem accounts for another quarter of all lamb losses.

Resistance to infectious disease

In the adult sheep, resistance to many diseases caused by agents such as bacteria and viruses is acquired by previous exposure to the agent. This previous exposure may be the disease itself, or treatment with a vaccine such as a clostridial vaccine, which induces resistance to a disease without actually causing it. Vaccination has two effects on the body's immune system. First, the production of antibodies which are found in the blood and elsewhere is stimulated. If infection occurs later these antibodies 'attack' the disease agents or their products and make them harmless. Second, the immune system is 'primed' so that when infection does occur, more of the appropriate antibody is quickly produced. With many vaccines the initial course of treatment consists of two injections with an interval between them. The first injection primes the immune system and stimulates the production of some antibody. The second injection stimulates the already primed immune system to produce more antibody. After the initial course of injections only a single booster injection is required periodically to stimulate the production of more antibody.

The newborn lamb has a problem. It has experienced neither disease nor vaccination. The antibodies in the ewe's blood cannot pass to the foetus (the developing lamb in the uterus) and thus vaccination of the ewe confers no immunity on the lamb before birth. (The same situation exists in the cow but in some species, such as man, antibodies can pass from the mother to the foetus before birth.) While antibodies in the ewe's blood cannot cross the placenta to the foetus they do cross into the udder and are concentrated in the colostrum (first milk). When the lamb sucks colostrum the antibodies are absorbed through the wall of the small intestine and enter the lamb's blood. The benefits of vaccination in the ewe are thus passed on to the lamb. But this benefit will only be fully acquired if the lamb sucks plenty of colostrum as soon after birth as possible and throughout the first twelve hours of life. After this time the antibodies cannot be absorbed through the wall of the small intestine. Some of the antibodies in colostrum are active within the gut itself and thus disease such as enteritis is much less likely if a lamb receives adequate colostrum.

The antibodies obtained from colostrum slowly wane in the lamb's blood for the immune system of the lamb itself has not been primed to produce antibodies. The lamb must later be vaccinated if protection is to be continued. When for some reason a lamb does not receive colostrum, temporary protection against disease such as lamb dysentery can be provided by an injection of antiserum.

Colostrum is obviously of great benefit to the newborn lamb but it can only give protection against diseases which the ewe has previously experienced itself by either infection or vaccination. If the lamb becomes infected with a bacterium or virus which the ewe has not previously met, it will have little defence. For this reason a high standard of hygiene is an essential part of good lambing management.

In summary, when a lamb is born it has practically no defence against infectious disease. The sucking of plenty of colostrum in the first few hours of life goes a long way to remedy this situation. In spite of this the newborn lamb is much more susceptible to infectious disease than the adult sheep and management must be adjusted accordingly.

An extra route for infection

An increased susceptibility to infection is bad enough, but the newborn lamb also has an extra route by which infection may enter the body – the navel. To understand the significance of this route we must examine a most fascinating aspect of circulatory physiology – the circulation of the foetus and the newborn.

Figure 1.1 shows a diagrammatic representation of the adult circulation (in both sheep and man). Oxygenated blood is pumped from the left heart into the aorta, and round the body, where it supplies oxygen and takes up carbon dioxide. The exhausted blood is returned in the veins to the right heart.

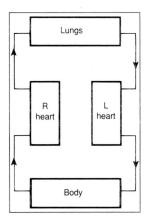

Figure. 1.1   Blood circulation in the adult

The right heart in turn pumps the blood through the lungs where it loses carbon dioxide and takes on more oxygen, and then to the left heart where the circuit begins again: a nice simple, efficient arrangement.

Affairs in the foetus are not quite so simple (Fig. 1.2). The components in Fig. 1.1, lungs, heart and body, are still there but we have gained the placenta, which for the foetus performs the function of the lungs, in addition to transferring nutrients from the ewe to the foetus.

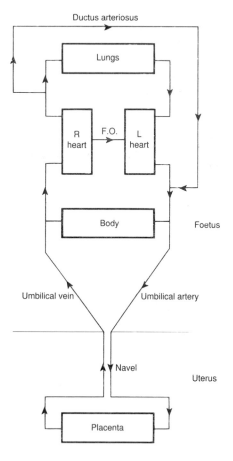

Figure 1.2   Blood circulation in the foetus. (F.O.-foramen ovale)

Let us now commence our circuit, as before, in the left heart. 40 per cent of the left heart output perfuses the body, but 60 per cent leaves the body via the umbilical artery and the navel to perfuse the placenta. Blood returns from the placenta via the navel and the umbilical vein to join the blood returning from the body. This mixed blood returns to the right heart.

The lungs in the foetus are not functioning. They are simply growing. Thus the requirement of the lungs for blood supply is much less than in the breathing lamb or adult sheep.

A reduced blood supply is achieved by means of two short circuits (Fig. 1.2), the foramen ovale (round hole) and the ductus arteriosus. The foramen ovale is in fact a hole connecting the right and left sides of the heart. If the foramen ovale persists into infancy, it is known as a 'hole in the heart'.

The effect of these short circuits is that approximately only 10 per cent of the blood leaving the right heart perfuses the foetal lungs; 50 per cent passes through the foramen ovale and 40 per cent through the ductus arteriosus.

Our circuit is now complete.

What happens at birth? The dramatic scenario which we see is accompanied by equally dramatic unseen changes in the circulation.

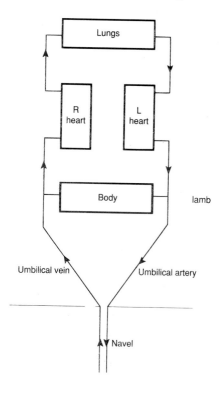

Figure 1.3   Blood circulation in the newborn lamb

As the lamb falls to the floor the navel cord breaks. This tearing action is vital, for in response the umbilical vessels contract and constrict, preventing haemorrhage. Cutting the cord eliminates this response and a fatal haemorrhage can occur.

Breaking of the umbilical vessels cuts off the placental supply of oxygen to the lamb, and very soon a shortage of oxygen stimulates breathing. As the first breath is taken the foramen ovale and the ductus arteriosus functionally shut and the newborn lamb suddenly acquires the circulation of the adult (Fig. 1.3).

In a premature lamb, (p. 79) the lungs may not fully expand. If this occurs the foramen ovale and the ductus arteriosus may not shut properly. The premature lamb may thus be handicapped by poor lung function and a leaky circulation.

What is the significance of this remarkable sequence of events to infectious disease? Although the umbilical artery and umbilical vein in the navel are functionally shut, they are not anatomically shut for some time. The umbilical artery leads to the aorta (the major artery of the body), whilst the umbilical vein leads, via the liver, to the vena cava (the major vein of the body). Thus there is a potential route for

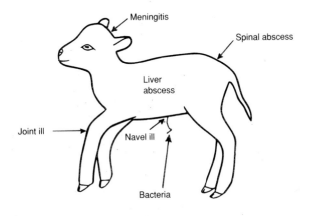

Figure 1.4   Diseases that can result from the entry of bacteria via the navel

infection to gain entry via the umbilical vessels into the general circulation, and then to lodge anywhere in the body. Figure 1.4 illustrates the common consequences of such infection.

Prevention of navel infection depends on three aspects of good management. First keep the lambing environment as clean as possible: plentiful new bedding is a minimum. Secondly, dress navels promptly after birth (see Chapter 7). Thirdly, ensure lambs receive adequate colostrum in the first hour of life. This last point is crucial to the prevention of all disease in newborn lambs.

In problem situations, preventative antibiotics may be required. This, however, must be regarded as a last resort.

Constraints on lamb viability

In spite of the inherent problems of the 'perfect' lamb, most such lambs survive and losses are much more likely to occur in lambs in which viability has for some reason been reduced. Many of the factors which reduce viability are largely related to events during the period of pregnancy – before the lamb is born.

Life starts at conception with the fertilisation of one or more eggs (ova) by sperm. For the next two weeks the embryo (fertilised egg) develops without any attachment to the ewe's uterus (womb). It

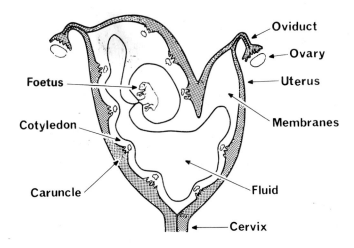

Figure 1.5   A schematic representation of a uterus 40 days after conception (adapted from Hunter, 1980)

receives its nourishment and oxygen from the fluid in the uterus. During the third week the developing embryo becomes attached to the wall of the uterus and the placenta starts to develop. We recognise the placenta at birth as the 'afterbirth' or 'cleansing'. This organ, which is part of the foetus, as the embryo is now called, attaches to the wall of the uterus and serves to carry food and oxygen to the foetus and to carry waste products in the reverse direction. In the sheep (and the cow) the attachments are at specific sites. On the wall of the uterus are small button-like structures called caruncles (this word literally means a fleshy lump). When the placenta contacts a caruncle, a corresponding cup-like structure is formed in the placenta – the cotyledon (derived from the Greek word *cotyle* meaning a cup) (Fig. 1.5; colour plates 1 and 2.) The cotyledons are the raised structures which we see in the afterbirth. It is through these units, each consisting of a caruncle and a cotyledon, that nourishment passes from the ewe to the foetus. From weeks 4–10 the major development in the uterus is the growth of the placenta. Comparatively little foetal growth takes place – a single lamb destined to weigh 5 kg (10 lb) at birth may weigh less than 0.5 kg (1 lb) at mid-pregnancy. The second half of a pregnancy is devoted to foetal growth during which time foetal weight may increase more than tenfold.

With this background we shall now consider some of the specific factors which can affect the development of the foetus and in turn the viability of the newborn lamb.

Placental constraints

The major determinant of foetal development and eventual newborn lamb weight is placental size. If only a moderate restriction is imposed a normal lamb will be born but it will be small, have low body energy reserves and may be born prematurely. If a more severe restriction is imposed the foetus will be short of oxygen in addition to food and a very small weak lamb will be born prematurely. If the placenta is very small the foetus may die and be stillborn – a foetal stillbirth (Fig. 1.6).

There are two factors which can influence placental size: litter size and ewe nutrition.

Placental development and litter size

The placenta is a part of the growing foetus and when there are two foetuses (twins) there will be two placentae. Triplets will have three placentae and so on. So far so good – but there is a problem. The number of caruncles (buttons) on the wall of the uterus is limited and

Figure 1.6   A newborn lamb with its stillborn twin. Note: this situation can be caused by infectious disease, such as toxoplasmosis

when there is more than one foetus the number of caruncles for each foetus is reduced. The size of each placenta is reduced and so therefore is its capacity to transfer nutrients from the ewe to the foetus. It is not surprising therefore that twins tend to be smaller than single lambs and triplets smaller than twins. This problem of smaller placental size can be especially serious when the two or three foetuses do not divide the available caruncles equally between them, leaving one foetus with a very small placenta indeed. The runt lamb in a set of triplets is a consequence of this situation.

Occasionally a ewe in good condition may produce a pair of small twins. Recent research has revealed one explanation for this. The 'twins' are not twin lambs at all – they are triplets! The 'missing' triplet died early in the pregnancy but after the distribution of the uterine caruncles between the three foetuses had taken place. The remaining two foetuses were unable to expand into the area of uterus occupied by their former mate and thus for the remainder of the pregnancy they received the nutrient supply appropriate to a triplet even though there were only two of them. Foetal death in early pregnancy is often associated with rough handling or poor nutrition.

In summary, twins and triplets are likely to be smaller and less well developed than single lambs because they have smaller placentae which restrict the passage of nutrients to the foetuses in the second half of pregnancy.

Placental size and ewe nutrition

It has become clear over recent years that ewe nutrition in the first half of pregnancy can affect the growth of the placenta and thus the later development of the foetus.

However the relationship between ewe nutrition and placental growth does not seem to be a simple 'more food – more placenta' one.

The ideal situation would appear to be a nutritional level before and during tupping to produce ewes (in a lowland flock) in condition score 3½ (see p.136). After tupping the level of nutrition should be continued to maintain body condition, or to permit a gradual loss of condition of half a condition score (but no more) over the first half of pregnancy.

What must be avoided is any abrupt change to nutritional plane, either up or down. If ewes are moved to flush grass for tupping, perhaps supplemented by concentrates, and then at the end of tupping are returned to bare pasture, detrimental effects on placental growth and subsequent lamb growth can be expected. The traditional 'stuff 'em, starve 'em, stuff 'em' method of ewe nutrition is not to be recommended.

Ewe nutrition

We have summarised the effects of nutrition during pregnancy in Table 1.2. Poor nutrition can have very different effects depending on its timing. The significance of poor placental development can clearly be seen. The effects of poor nutrition in the first half of pregnancy cannot be fully overcome by improved nutrition in the second half. By contrast poor nutrition in the second half of pregnancy following good nutrition in the first half has minimal effects on lamb growth, but disastrous effects on ewe condition and colostrum production.

In addition to the effects noted in Table 1.2 poor nutrition at tupping will result in a low ovulation rate and thus fewer lambs conceived. It will also result in an increased rate of embryonic mortality, death of fertilised eggs before they become attached to the wall of the uterus.

Table 1.2.   The effects of nutrition during pregnancy

Nutrition		Effects on:				
First half	Second half	Placental growth	Foetal growth	Birth	Ewe condition at lambing	Colostrum Supply
Good	Good	Good	Good	Normal	Good	Good
Poor	Poor	Poor	Poor	Premature	Thin	Little or none
Poor	Good	Poor	Poor	May be premature	Fair	Fair
Good	Poor	Good	Moderate	May be premature	Thin	Little or none

Ewe age

Mortality is generally higher in lambs out of either very young or very old ewes.

The young ewe is an inexperienced mother. She takes longer to lick her lamb dry and may be unwilling to stand for sucking. These problems are most evident when twins are produced. The older ewe by contrast tends to be a good mother but poor nutrition, often related to teeth or feet problems, can lead to the birth of small weak lambs and a shortage of milk.

Congenital abnormalities

A congenital abnormality is any abnormality present at birth. These abnormalities result from some interference to the development of the foetus during pregnancy. The problem may be inherited from one of the lamb's parents. Entropion (turning-in of the lower eyelid) would seem to be one example since it is more common in some breeds than in others. Other conditions are not inherited and result from some 'outside' interference. Swayback is one example in lambs (see Chapter 3) and the thalidomide disaster in children is a tragic example from the human world.

Many congenital abnormalities in lambs do not threaten life directly but may do so indirectly. Entropion, if not treated, leads to blindness and thence starvation. Deformities of the jaw can have the same result since they often make sucking either difficult or impossible.

Birth

Two aspects of birth can markedly affect viability. The first is prematurity and the second is hypoxia (a shortage of oxygen) during birth itself.

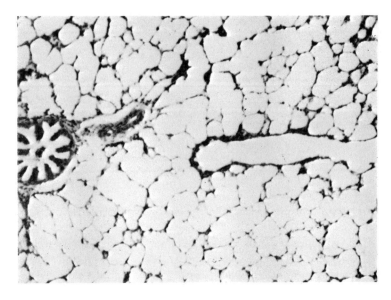

Figure 1.7   A histological section of lung tissue from a healthy newborn lamb. The lung has fully expanded. (Picture by J. S. Gilmour)

Premature birth is associated with poor ewe nutrition, and with twin or triplet litters. It may also be caused by infectious disease such as enzootic abortion. In all cases the result is the birth of small, weak lambs of low viability. The more premature the birth, the greater the problem. Premature lambs have poor birth coats and a low capacity to produce heat and are thus very susceptible to hypothermia. These lambs are physically weak and may not be able to suck. Even if they can suck they often go hungry, since ewes which lamb prematurely often have no colostrum. Premature lambs can have breathing problems because the lungs may not fully expand when the first breath is taken (Figs. 1.7 and 1.8). With careful nursing many premature lambs will survive, but this is a time-consuming and often frustrating exercise and prevention is much better than cure.

Severe hypoxia during birth is a problem which probably affects about 3 per cent of all lambs born. During pregnancy, and most of the birth process, the lamb derives its oxygen supply from the ewe via the placenta. Immediately the lamb is born the task of supplying oxygen is taken over by the lungs when the lamb starts to breathe. Inevitably in many births there is a gap between stopping the placental supply of

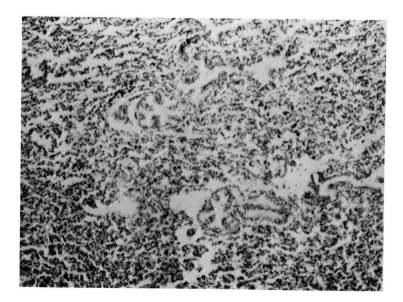

Figure 1.8 A histological section of lung tissue taken from a premature lamb. The lung has not expanded. (Picture by J. S. Gilmour)

oxygen and starting to breathe. Provided this gap is fairly short no problems arise. But if the gap is unduly prolonged, as might occur in a difficult lambing, the lamb may die from hypoxia (shortage of oxygen) and be stillborn – a parturient stillbirth. Some lambs affected by hypoxia do survive, but only just. These lambs appear lifeless after birth and quickly become hypothermic for they can produce very little heat.

This lifeless state is thought to be caused by an acidity (low pH) of the lamb's blood, a product of the hypoxic period. This condition, known as metabolic acidosis, is self-correcting provided hypothermia is prevented. Affected lambs should be dried and placed in a warmer for the first few hours of life (see Chapter 7).

Summary

1. All lambs are disadvantaged when compared with adult sheep because:

    (a) they have a high energy requirement but only have low energy reserves and are totally dependent on the ewe for food;

(b)　they have poor coats and are born wet;

(c)　they have little resistance to disease; this problem is overcome to a large extent when the lamb sucks plenty of colostrum.

2.　Viability is further restricted if:

　　(a)　the lambs are twins or triplets;

　　(b)　ewe nutrition is poor;

　　(c)　the ewe is either very young or very old;

　　(d)　the lamb is affected by a congenital abnormality;

　　(e)　the lamb is born prematurely;

　　(f)　the lamb suffers severe hypoxia during birth.

By now you may have the impression that most lambs are born with a death wish. They are not! The notes above do tell us something positive:

1.　Many problems can be avoided by good management both before and at lambing.

2.　We know which lambs are likely to require most attention at lambing. A single out of a fit ewe will do very well without our interference, whereas triplets out of an old ewe or twins out of a ewe-lamb will probably benefit from a little human help.

Chapter 2

Lambing the ewe

Much has been written over the years on assisting the lambing ewe but still many lambs and ewes die needlessly. This is not through any lack of effort by lambing shepherds but is rather a case of doing the wrong thing at the wrong time and, last but not least, not knowing when to stop and seek professional assistance. In the notes below we have described the common problems met at lambing and how these should be approached. Throughout we have tried to indicate when the shepherd should stop and summon professional help.

The normal lambing

Signs of lambing may be seen some time before the birth actually begins. The ewe may not come to the feed trough and may separate herself from the flock. If closely watched she may be seen to periodically lift up her head and purse her lips – her uterus (womb) is contracting. These uterine contractions progressively become more frequent until the ewe starts to strain and bear down, and it is obvious that something is happening.

The first physical sign of lambing at the vulva varies from ewe to ewe. The 'water bag' (fluid filled membranes) may be ejected and hang from the vulva or in some cases the bag may burst within the ewe and only fluid be ejected. In some cases the first observed sign may be part of the unborn lamb. In a normal presentation (Fig. 2.1) the forefeet appear first with the head just behind. The ewe may take anything from a few minutes to half an hour to complete the delivery. Older ewes are generally quicker, as are ewes having twins or triplets. If a ewe does have twins or triplets the second lamb may be delivered within minutes of the first but in many cases contractions cease and the delivery of the next lamb may occur after a delay of up to an hour. This delay has some advantages for it gives the ewe time to lick the first lamb dry.

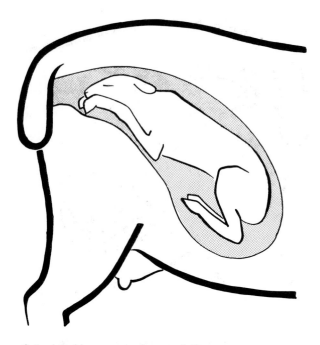

Figure 2.1   Lambing – normal presentation

Most of us have assumed that the normal lambing position was the 'diving' position, front legs fully extended (Fig. 2.2). Recent research has shown that this is not the case. The normal position involves flexion of both elbow and shoulder (Fig. 2.1).

In an unassisted birth this flexed position presents no problems but if a birth is to be assisted by applying traction to the front legs the position must first be changed to the 'diving' position, front legs extended.

To straighten the leg, cup the elbow joint with one hand and pull the leg with your other hand. Once the legs are straight the ewe can easily be lambed.

Approach to assistance

When to interfere

There are a number of circumstances in which help is clearly required. These include:

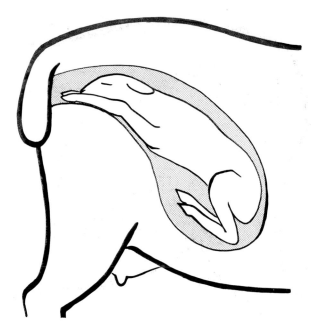

Figure 2.2   Lambing – forelegs extended; position necessary for assisted birth

1. Only the head appears.
2. The water bag has been delivered or has burst and there has been no progress for 30 minutes.
3. The total period of lambing has exceeded 90 minutes.
4. A tail or only one leg has been delivered.

But on many occasions the situation is not so clear-cut. If in doubt the ewe must be examined to check that all is well, not necessarily to deliver the lamb. If the lamb is alive (see below) and in normal presentation (Fig. 2.1) she may be left for another 30 minutes. A forced lambing before the birth canal is fully open is at least very painful for the ewe and at worst may cause her death. If however the lamb is dead, or incorrectly presented, help is needed.

Hygiene

Be as clean as you possibly can. Poor hygiene at lambing leads to metritis (a serious infection of the uterus, see Chapter 4) and dead ewes. The area around the vulva should be dagged if this has not already been done and the whole area washed with soap and water containing a non-irritant disinfectant. Placing a clean paper sack under the ewe's hindquarters helps to keep the working area clean. Thoroughly wash your hands and arms and remember to keep nails well trimmed. If at all possible have an assistant hold the ewe. This not only helps you to keep clean, you will also be more gentle.

Lubrication

Good lubrication is essential if damage to the ewe is to be avoided. A number of lubricant creams, gels, oils and powders are available.

Gentleness

Be gentle at all times. Force rarely achieves results, and damage to the ewe and her death is the likely sequel. Remember to remove any rings from your fingers.

Ewe position

This is a matter of personal preference but it often helps to position the ewe so that the lamb's offending limb or head is uppermost.

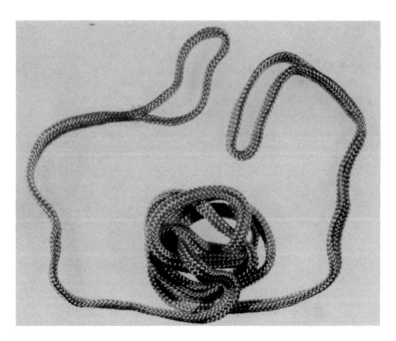

Figure 2.3   Soft lambing rope fitted with loops at both ends

Occasionally it can help to raise the ewe's hindquarters off the ground. This takes the weight and pressure of the abdominal contents off the uterus and may ease the correction of a malpresentation. Only keep the ewe in this position for a short time as it will make it difficult for her to breathe.

Retropulsion

All malpresentations are much easier to correct if the lamb is first pushed back into the uterus – retropulsion. This is best achieved by pushing steadily on the lamb for a few seconds. Do not use excessive force.

Identification of legs

It will often be important to decide on feel alone whether a leg is a foreleg or a hind leg. This is simple providing that you have practised beforehand. The knee joint of the front leg feels very different from the hock joint of the hind leg. Working your fingers up a foreleg you will feel the sharp spine of the shoulder blade. Working your fingers up a hind leg will bring you to the pelvis and behind that the tail.

It is equally important to be able to check that the two legs that you have found are a pair belonging to the one lamb. First check that they are a pair: two forelegs or two hind legs. If the legs do belong to one lamb you should be able to run your fingers up one leg and down the other without losing contact with the lamb.

Is the lamb alive?

In some cases it may be obvious that the lamb is dead. The fleece or even parts of the lamb may come away in your fingers. If unsure place your finger in the lamb's mouth. A healthy live lamb will respond with a suck. If no suck is felt the lamb is either very weak and in serious trouble, or dead.

Lambing aids

Two aids can be of great benefit, ropes and a snare.

The traditional lambing rope is baler twine but soft 'snakeskin' ropes fitted with loops at both ends (Fig. 2.3) are much easier to use and are kinder to the lamb's legs.

Ropes can be most profitably used for securing a leg which has to be replaced to enable a lamb to be repositioned. Once the manipulation has been achieved the leg can easily be regained.

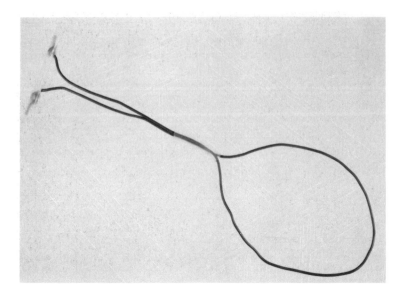

Figure 2.4   Lambing snare

The snare (Fig. 2.4) is a most useful piece of equipment. Its use is fully described on page 24. The snare must be positioned behind the lamb's ears (Fig. 2.5). If it is not it will fall off.

Time

Most successful shepherd interference is completed within five minutes of starting. If you have made no significant progress within ten minutes you are probably stuck! Unless you are very experienced and know exactly what you are doing stop and get help from your veterinary surgeon. Try to avoid the sequence of student–tractor driver–shepherd–foreman–farm manager–vet! This leads to dead lambs, dead ewes and frustrated vets!

Using your vet

It is most helpful if you discuss the management of lambing problems with your veterinary surgeon before lambing starts. In many cases it will be preferable for you to take the ewe to the surgery.

Figure 2.5   Snare and ropes positioned on a lamb. Snare must be behind the ears

Training

If you have not done so already, go on an Agricultural Training Board lambing course.

Specific problems

Normal presentation (Fig. 2.1)

If your examination reveals a normal presentation, leave the ewe for a further 30 minutes. If after this time there is no further progress ensure that the birth canal is well lubricated, check that each leg is straight and pull the lamb towards the ewe's hind feet. It often helps to pull the legs alternately and to slightly rotate the head. If this is of no avail it is likely that the lamb is very big and your vet's help is needed.

Front leg(s) back (Fig 2.6)

This is probably the most common malpresentation. One or both legs may be turned back. Draw the offending limb into the correct position

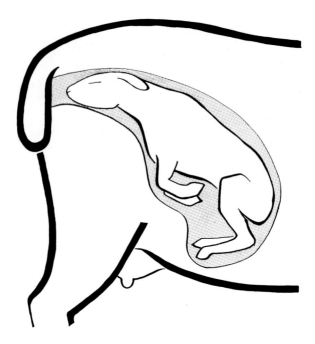

Figure 2.6   Lambing – two forelegs back (bilateral shoulder flexion)

by cupping the foot in your hand to protect the wall of the uterus. In the case of a second twin or a comparatively small lamb it may be possible to deliver the lamb with one leg back. This should not be attempted with a big lamb, especially if out of a maiden ewe.

Head back (Fig 2.7)

One or two forelegs are presented but the head is deflected. This can be a most difficult problem to resolve. It is often possible to retrieve the head by cupping it in your hand but the head is lost each time delivery of the lamb is attempted. This problem is best corrected using the snare. Apply the snare behind the lamb's ears (Fig. 2.5) and secure both forelegs in the birth canal with ropes. Apply gentle traction to the snare and guide the lamb's nose towards the birth canal. Once the nose is within the pelvis the head cannot be lost. Now gentle traction on the snare and the ropes should achieve delivery.

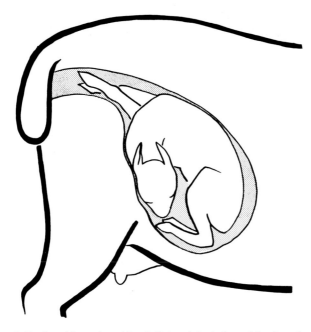

Figure 2.7   Lambing – head back (lateral deviation of the head)

Posterior presentation (Fig 2.8)

This is simply a case of the lamb coming backwards and is quite common in multiple births. It is inadvisable to attempt to reverse the lamb's position. Deliver the lamb in this position ensuring plenty of lubrication. Sometimes a lamb in posterior presentation is also upside down, i.e. the lamb's belly is nearest to the ewe's back. Deliver this lamb by pulling straight out of the birth canal – not towards the ewe's feet.

Breech presentation (Fig 2.9)

This term describes a lamb in posterior presentation but with the hind legs pointing towards the ewe's head. Using plenty of lubrication, each hind foot is cupped in the palm and brought into the birth canal. This sounds quite simple but if it is a large lamb it may be extremely difficult. Seek help if you are in doubt.

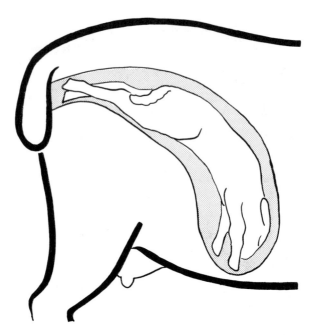

Figure 2.8   Lambing – lamb coming backwards, hind feet first
(posterior presentation)

Twins (Fig. 2.10)

Twins present no specific problems providing you are sure which legs
belong to which lamb.

Ewe-lambs and gimmers

Ewe-lambs and gimmers present special problems. There is less room
for manoeuvre in these ewes and the correction of malpresentations
tends to be more difficult. This does not mean that these ewes should
be assisted in lambing at the earliest possible moment. To do so would
be harmful, for these ewes need longer to lamb to enable the birth canal
to open fully. It does mean, however, that these ewes should be
checked at an earlier stage to ensure that everything is correct. Two
forelegs back in a ewe-lamb is easier to correct before the head has
come out. After the head has come out it can be almost impossible.
Lambing problems in ewe-lambs and gimmers are often associated
with overfatness. This should be avoided.

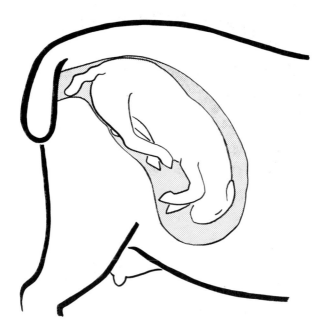

Figure 2.9   Lambing – breech presentation. The hind legs are pointing forwards

Dead lambs

Dead lambs present the lambing shepherd with some difficult problems, especially if decomposition has set in. The ewe's lambing efforts may be half-hearted and the contents of the uterus are likely to be dry, making progress slow and difficult. Copious lubrication is required together with absolute gentleness since the wall of the uterus may be easily damaged. If at all in doubt contact your veterinary surgeon.

Ringwomb

In this condition the entrance to the uterus, the cervix, fails to dilate although it is clear that the lambing process has started. A small hole, the size of a ring (hence the name) is felt. Check the ewe 30 minutes and one hour later. If there is no change contact your vet. A Caesarian section may be required. Ringwomb is sometimes confused with an incomplete relaxation and opening of the birth canal. In this condition

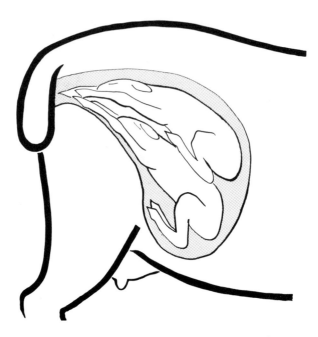

Figure 2.10   Lambing – twins. Both lambs are in normal presentation

the hand can be introduced and the lamb felt but there is insufficient room for delivery. This is most common in fat ewes which often appear to give up the lambing effort. Clenching of the fist within the birth canal may help in some cases but unless swift progress is made you need veterinary help.

Uterine inertia

This is a comparatively rare problem in sheep. It literally means that the uterus apparently makes no expulsive efforts.

Inertia can occur in two situations. In the first, primary inertia, the uterus fails to contract at any time. The cervix is often found to be open and lambs can be felt, but the uterus seems to be making no effort to expel its load.

The second type of inertia, secondary inertia, occurs after the uterus has made prolonged but unsuccessful efforts to expel the lamb. The problem in this case is simple fatigue.

Whatever the type of inertia the procedure is the same. Observing all the precautions outlined, especially copious lubrication, attempt to deliver the lambs.

Do not persevere too long, ten minutes is quite enough. The survival of the lambs is at considerable risk. If your efforts prove unsuccessful take the ewe to your veterinary surgeon. A Caesarian section may be required.

Uterine torsion

This is another relatively unusual event but one to which shepherds should be alert.

Uterine torsion describes a twist at lambing of the uterus within the body, the twist being around the cervix and the vagina. This inevitably blocks the exit of the lambs from the uterus.

The ewe enters first-stage labour, uterine contractions only, but makes no progress. The observant shepherd will notice this ewe and when progress is clearly not being made, will examine her. This examination will reveal the problem. The birth canal is totally blocked and your hand feels as though it is being turned as it enters the torsion.

This is an emergency. If remedial action is not quickly taken the lives of the ewe and the lambs are at risk.

Attempts can be made to correct the twist by rolling the ewe on her back from side to side – roll the body in the opposite direction to the twist. A hand held in the birth canal will reveal if you are successful. The aim of this exercise is to roll the body around the uterus. Under no circumstances should any attempt be made to roll the uterus within the body by internal manipulation.

If successful, lambing may proceed normally but keep a careful watch on the ewe. The cervix may fail to dilate properly and veterinary help may be required.

If your rolling efforts are in vain take the ewe to your veterinary surgeon. A Caesarian section is required.

Resuscitation of the newborn

One of the most frustrating situations faced by us all at lambing is the newly born live lamb which obstinately refuses to breathe. Some of these lambs will never breathe but some can be saved by resuscitation.

There are a number of causes of failure to breathe. The commonest is prematurity (p. 79). In these lambs, the lungs are not fully mature and expansion may be impossible, or only partial expansion may be

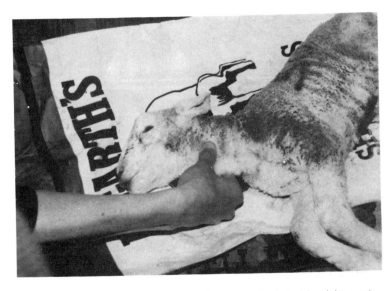

Figure 2.11   Occlusion of the oesophagus on the left side of the neck prior to resuscitation

achieved. Resuscitation may have beneficial effects in some of these lambs.

Blockage of the airways with mucus may cause suffocation in some lambs. If the blockage can be relieved, resuscitation may well be effective.

Some lambs suffer hypoxia (a shortage of oxygen) during birth and are born in a very weak state. They are too weak to breathe. Resuscitation will hopefully benefit these lambs. Bear in mind, if resuscitation is successful, that such lambs constitute high hypothermia risks.

A few lambs are born fit and healthy but seem a little lazy in commencing to breathe. The first breath may only be taken a minute after birth. Inevitably resuscitation in these lambs is very successful!

For practical use we need a resuscitation regime which we can apply to all problem lambs. This regime must hopefully achieve two aims: (1) removal of any obstruction and (2) inflation of the lungs.

Removal of any obstruction is easier said than done. A number of suction devices are available and may be useful. They will remove mucus from the throat but not from the trachea (windpipe). The

Figure 2.12   Resuscitation of a newborn lamb which has failed to breathe

traditional swing is always worth a try but don't overdo it. Once round is quite enough.

In medical practice the paediatrician would pass a special tube into a weak baby's trachea. Once in place the lungs are inflated via this tube. This is practically impossible in the lambing shed but there is an acceptable practical alternative.

1. Place the lamb on its right side and occlude the oesophagus (gullet) by applying your right thumb to the left side of the neck (Fig. 2.11). If this is not done your attempts to inflate the lungs will simply inflate the stomach.
2. Then place the first two inches of the stomach tube into the lamb's mouth. Occlude the lamb's nostrils and seal the stomach tube with your left hand.
3. Now blow smartly but briefly down the stomach tube (Fig. 2.12). Relax and repeat at two second intervals.
4. If successful you will be rewarded by expansion movements of the chest wall. Continue until the lamb commences to breathe.
5. If you do not see chest wall movements the outlook is poor. Continue your efforts for a minute and then give up.

This is a simple technique but it does require a little manual dexterity. Practise on a dead lamb until you get the hang of it.

This technique inevitably involves close contact with the lamb. Place the tube in your mouth before touching the lamb. If you suspect infection, e.g. infectious abortion (see Chapter 5), DO NOT use this technique.

After an assisted lambing

Ensure that ALL the lambs have been born. Administer long-acting antibiotic by injection to the ewe (consult your veterinary surgeon about this before lambing starts). Check that the ewe has not suffered physical damage – if she has, take professional advice. Keep an eye on the ewe to check that the placenta (cleansing, afterbirth) comes away cleanly. Retention of the placenta, which can lead to metritis, is more common after an assisted lambing. If the ewe is fit and strong, place the lambs at her head and leave them alone. If the ewe is weak, the lambs will need drying and feeding to prevent hypothermia.

Don't forget:
1. Be gentle – do not use force.
2. Use copious lubrication.
3. Be as clean as you possibly can.
4. Do not hesitate to summon professional help.

Chapter 3

Problems in lambs

In this chapter we have listed alphabetically the common problems of lambs. Brief notes on causes, symptoms, treatment and prevention are provided. When appropriate the reader is referred to a page in the relevant section in Chapter 7 (Techniques for treating newborn lambs). The need for brevity inevitably means that much detail has been omitted. We would encourage readers to consult their own veterinary surgeon and also the literature included in the further reading list (p. 201) for more information.

Before lambing the shepherd should consult his own veterinary surgeon for guidance on the treatment of the conditions likely to occur in his flock. Routine treatments can be discussed and criteria for deciding when professional help is needed established.

Abscess
See Liver abscess *and* Spinal abscess.

Acidosis
See Ruminal acidosis.

Arthritis
See Erysipelothrix infection.

Atresia ani
This is a congenital condition in which the opening of the back passage – the anus – is missing. In some cases only a thin membrane blocks the

anus, and if this is perforated, all is well. In more serious cases the rear portion of the hind gut is missing.

Symptoms
Initially the lamb appears perfectly normal. After a day or so the abdomen enlarges and straining may be seen. A close examination reveals the problem. If only a membrane is blocking the anus a swelling may be found in the anal region caused by the accumulation of faeces in the bowel.

Treatment
This is most definitely for your vet.

Prevention
The precise origin of this congenital defect is not clear. There is no cause for concern providing only the occasional lamb is affected.

Atypical pneumonia
Atypical pneumonia occurs in lambs aged more than two months, though in some circumstances lambs aged only three weeks can be affected. Many lambs in a flock can be infected but mortality is generally low. This disease is caused by infection with both mycoplasma (small bacteria) and the bacterium *Pasteurella haemolytica*.

To a considerable extent, atypical pneumonia is a disease of intensive husbandry, being practically unknown in lambs on the hill, but relatively common in intensive, inside finishing systems.

Symptoms
Chronic coughing is the common sign, especially when lambs are disturbed or driven. Many lambs remain bright but some may appear depressed. Growth rate is likely to be reduced. The occasional death may occur.

Treatment
Atypical pneumonia can be treated with antibiotics, but these tend to depress the infection rather than eliminate it. This treatment is probably inappropriate in lambs which appear bright, only exhibiting the occasional cough. Treatment is indicated in lambs which appear depressed by the infection.

Prevention

Vaccination and antibiotics have no part to play in preventing this condition. The approach must be a management one. In housed lambs, careful attention must be paid to stocking rate and ventilation. The aim must be to reduce the weight of infection within the lambs' environment.

In intensive feeding systems it is preferable to avoid mixing batches of lambs, at least until the lambs have adapted to new feed and a new environment. Hill lambs are especially unlikely to have any resistance to this infection and mixing these naive lambs with infected stock is a recipe for disaster.

Bladder stone

See Urinary calculi.

Bloat

Bloat is a relatively unusual problem in sheep unless the animal is cast on its back and is unable to eructate (belch). However it can occur especially in young, artificially reared lambs, normally aged under one month.

Symptoms

The abdomen is considerably distended. The condition in young lambs often tends to be more chronic than in adult ruminants, and the lambs may well continue to walk around. Loss of appetite is common.

Treatment

In the adult sheep or the ruminating lamb (aged more than three months) the problem is ruminal bloat. Emergency treatment with a trocar and cannula (or failing this a sharp knife) in the left flank will relieve acute cases. Soap powder by mouth is an effective treatment.

However, in young lambs dilation of the abomasum (fourth stomach) is the problem. Do not attempt surgical treatment. Contamination of the peritoneal cavity (body cavity) with abomasal contents will likely lead to a fatal peritonitis.

In chronic cases in young lambs, antibiotic treatment is indicated to control the fermentation in the abomasum which is causing gas production. Antibiotics must be given by injection; agents given by mouth will probably not reach the abomasum. In acute cases seek urgent veterinary assistance.

Prevention

Abomasal bloat in young lambs is related to feeding practice. Twice daily bottle feeding with large quantities of warm milk, especially from dirty equipment, predisposes to this condition. The problem is exacerbated by the hunger which naturally develops between feeds. The lambs pick around the pen consuming everything from bits of wood to wool and baler twine. These indigestible objects lodge in the abomasum.

Abomasal bloat is rarely seen in lambs fed on ad lib cold milk from clean feeding equipment (p.177).

It is sometimes recommended that formalin (37 per cent formaldehyde) is added to the milk at the rate of 0.1 per cent to prevent this problem.

Blowfly strike

This must be one of the most distressing diseases of sheep, for both man and beast! Open wounds and fleece contaminated with faeces attract the attention of flies such as the black blowfly, the bluebottle or the greenbottle. These flies lay their eggs which quickly hatch into larvae. These larvae (maggots) then proceed to attack the skin and literally eat the sheep.

Symptoms

On examination the condition is obvious, an open foul smelling wound with maggots in evidence.

The detection of a struck ewe or lamb from a distance is not quite so simple, but it is an art which must be learnt. An experienced shepherd will detect a struck animal from a distance exceeding 100 yards.

The struck animal is restless, behaving in an unusual way. It may attempt to kick or bite the affected part. Grazing is much reduced.

Treatment

Once detected the struck sheep must be promptly treated. Clip the surrounding wool and remove as many maggots as possible – a perfect job for your student apprentice! Then apply a larvicidal ointment or powder. Antibiotics by injection may well be indicated to control secondary bacterial infection. Keep the sheep indoors until the wound is fully healed. Remember the rest of the flock will be at risk.

Prevention

There are two approaches to prevention. First avoid the factors which attract the flies. In lambs, scouring due to parasites is a major problem. Control the parasites (p.74). If fleece soiling is present, remove the wool in the contaminated area (crutching).

Wounds should be avoided at all times, but especially when flies are active. If wounds do occur, and there is a risk of blowfly, treat the wounds with a larvicidal ointment and check the animals over the next few days.

The second approach is to use an insecticide either as a dip or a pour-on preparation. This treatment must be given before the anticipated risk period.

The use of insecticides is an additional means of prevention, not an alternative to good husbandry, e.g. controlling parasites and treating wounds.

Bone abnormalities

See Rickets.

Border disease

Border disease is a virus infection. The ewe initially becomes infected but shows no signs of disease. The infection passes from the ewe to the foetus in the first half of pregnancy. This can cause abortion at about 90 days of pregnancy or can lead to the birth of weak deformed lambs. Infection occurs most commonly in young ewes – ewe-lambs and gimmers.

Symptoms

These are very variable. Abortion at around 90 days may be observed and an unusually large number of ewes may be barren at lambing. Affected live lambs may be merely weak or may show the characteristic 'hairy shaker' signs. The 'hairy shaker' lamb has a coarse birth coat and in smooth-coated breeds the coat may be pigmented. Tremors may be observed over the back and in the legs. The head may appear domed in shape and defects of the jaws and legs may be present.

Treatment

There is no specific treatment. Affected lambs may survive with careful nursing, but they seldom thrive.

Prevention

Consult your veterinary surgeon for an accurate diagnosis and assessment of the problem. Affected lambs should not be retained for breeding, and should be slaughtered at least one month before tupping begins (pp. 125–126).

Broken leg

See Fractures.

Castration (incorrect)

Occasionally an inexperienced operator fails to use correctly the rubber ring method of castration. Either only one testicle is included below the ring, pushing the other testicle high into the scrotum (purse), or the ring may be applied too high, interfering with the urethra (the tube connecting the bladder to the penis); see colour plates 11 and 12.

Symptoms

Both the problems referred to above result in pain and discomfort in excess of that normally associated with castration. A few hours after castration, when most lambs appear normal, affected lambs stand awkwardly with the hind legs apart. They are unwilling to walk.

Treatment

Remove the ring. This is made easier if a small blunt instrument such as a teaspoon handle is first passed between the ring and the skin. The ring may then be cut safely without risk of cutting the skin. If there is any doubt as to whether the lamb has sucked plenty of colostrum give tetanus antiserum (consult vet). Leave castration to another day – the lamb has had enough.

Prevention

Proper instruction (p. 185).

Cerebrocorticonecrosis (Polioencephalomalacia, CCN)

This is an acute nervous disease of sheep especially growing lambs, caused by a deficiency of the B vitamin, thiamine. In healthy sheep thiamine is produced by the ruminal bacteria but in some circumstances other ruminal bacteria produce the enzyme thiaminase, which destroys thiamine. The result is thiamine deficiency and acute disease.

Symptoms

If the early stages are detected, animals may scour and appear dejected. In many cases however the first sign of trouble is a recumbent sheep displaying nervous signs. These signs include trembling, blindness and paddling of the legs. The head may be held well back. Often only one or just a few animals are affected.

Treatment

Treatment is only successful if given at the earliest possible stage. The sheep should be injected with thiamine or a multi-vitamin preparation containing thiamine. An intravenous injection by your veterinary surgeon will achieve the best result. Recovering animals should receive careful nursing.

Treatment should never be delayed when cerebrocorticonecrosis is suspected. It will do no harm even if this disease is not the problem.

Prevention

The exact reason why some ruminal bacteria produce thiaminase is not known, but it is likely diet related. Thus it makes sense to change the diet, which will in turn cause a change in the ruminal bacterial population. This will often mean a move from lush to bare pasture, or withdrawal of concentrates.

Chilling

See Hypothermia.

Cleft palate

Cleft palate is a developmental defect in which the roof of the mouth is not properly formed. The result is a physical connection between the mouth and the nasal passages. Affected lambs cannot suck properly.

Symptoms

Starvation will probably be the first symptom seen. If the lamb is fed with a bottle, milk may be seen running out of the nose. The lamb will easily choke. An examination of the lamb's mouth with a finger will reveal the problem.

Treatment
None. Affected lambs should be humanely destroyed.

Prevention
Take professional advice if more than the occasional lamb is affected.

Cobalt deficiency
See Pine.

Coccidiosis
Coccidiosis is another disease of intensive husbandry. It is caused by infection of the intestine with a protozoan parasite called *Eimeria*. Lambs are most commonly affected at 4–6 weeks of age. Infection, which occurs in most lambs in lowland systems, leads to life-long immunity. A low-level infection produces no clinical signs of disease, but a heavy infection does produce serious disease.

Symptoms
Acute diarrhoea accompanied by dullness and abdominal pain with inappetence are the presenting signs. The scour can sometimes contain streaks of blood. The disease leads to dehydration and in severe cases death. Affected lambs rapidly lose flesh.

Treatment
A number of effective drugs are available. An injection of long-acting sulphonamide is probably the treatment of choice. In severely affected lambs the dehydration must be corrected by fluid given by stomach tube (p.159). Remember other apparently healthy lambs are at risk.

Prevention
Ewes carry a very low level of infection and produce a few infected oocysts (eggs). This low number of oocysts poses no immediate threat to lambs. Indeed, there is evidence to suggest that a low level of

infection in newborn lambs may help to stimulate immunity and prevent serious disease later in life.

In intensive situations, however, the low level of infection in lambs, derived from the ewes, can lead to problems. The lamb infected with a low dose of oocysts acts as a multiplier, passing out many more oocysts than it consumed. After a few cycles of this process the ground becomes heavily contaminated with oocysts. This presents a serious challenge to lambs, which will succumb to clinical disease.

This pattern of infection gives some clues in the management of lambs to prevent coccidiosis. When ewes and lambs leave the sheep house they should proceed in a radial manner, each group going to clean fresh pasture. Do not pass all the lambs through the same series of pastures: this is a recipe for coccidiosis. Move feeding troughs daily to avoid creating a heavily contaminated area.

If management measures are unsuccessful, some form of medication must be used. The method and drug used will depend on the rearing system. If lambs are to be fed from a young age, in-feed medication can be used. If lambs are not to be fed, strategic dosing with an anti-coccidial drug can be employed, dosing the lambs before the known risk period.

Drug treatment of ewes before and after lambing has been suggested to reduce their oocyst output. However, this low level of contamination may help newborn lambs to acquire some immunity to coccidiosis. The value of this technique is open to question.

Constipation

This is an unusual condition in newborn lambs. Failure to pass the first dung, the meconium, may occur in watery mouth but this is a consequence of this condition and not a cause. Constipation itself is most likely in lambs which have failed to suck properly and have been fed by stomach tube.

Symptoms
The lamb appears listless and may not suck. It may or may not strain. The anal region is clean.

Treatment
Give 5 ml liquid paraffin by mouth, safest by stomach tube (p. 159) – and administer an enema (p. 189).

Prevention
Lambs which suck well from birth are unlikely to suffer from this problem.

Copper deficiency
Copper deficiency in the ewe leads to swayback in lambs (p.88), but copper deficiency in growing lambs can have deleterious effects, even in the absence of swayback in a flock. The problem is most commonly seen in hill lambs grazing improved pastures.

Symptoms
Symptoms are vague and non-specific. They include a staring greyish coat, poor growth rate, anaemia (seen as paleness of the gums, conjunctiva or vulva), susceptibility to fractures and an increased susceptibility to infectious disease.

There is another problem in detecting this condition. In many diseases, the sick animal is obvious since it differs in behaviour and appearance from the rest of the flock. Not so in copper deficiency: all the lambs are similarly affected. The most likely complaint to the veterinary surgeon is that the lambs are not doing as well as was hoped.

Investigation by your veterinary surgeon is required. Flock treatment for copper deficiency is expensive and will be of no benefit if the wrong disease is treated. Samples must be taken for analysis to confirm copper deficiency, and to detect other problems such as cobalt deficiency and parasitic gastroenteritis.

Treatment and prevention
Copper can be given to the lambs either orally or by injection.

Copper oxide needles contained in capsules are the oral treatment of choice. The needles lodge in the abomasum where copper is slowly released. In situations where cobalt or selenium deficiency also exist, use of the 'Cosecure' glass bolus is indicated (p.148).

A number of injectable preparations are available. Some of these agents can cause a reaction at the injection site. In excess, copper is highly toxic (see Copper poisoning). Thus, only one form of copper supplementation should be used. Do not treat lambs if they are soon to be moved to a situation where dietary copper will be readily available, e.g. hill lambs being sold to intensive finishers.

Copper poisoning

Copper poisoning may be acute or chronic, and is normally related to a high copper content in the feed, unnecessary copper supplementation or both these factors.

Breed plays a part. Scottish Blackface sheep are relatively resistant to copper poisoning whereas Suffolks, Texels and many of the recently imported breeds are highly susceptible.

Symptoms

Acute poisoning often follows copper supplementation, especially injection, within 2–3 days. A dead sheep is the common sign.

In early chronic poisoning the signs are vague and non-specific, the non-doer, but eventually a crisis point is reached. The animal becomes acutely ill and death rapidly follows.

An accurate diagnosis must depend on detection of excess copper in tissue samples from the dead lamb and in blood samples from the live lamb.

Treatment

Treatment of the acutely ill sheep may well not be successful but detection of the problem indicates preventative treatment for similar stock. Injection and feeding of molybdenum compounds decreases the absorption of copper and increases its excretion. All concentrate feeding must stop.

Prevention

Copper supplementation must never be employed unless there is good indication for doing so (e.g. low blood copper values). To do otherwise is a waste of money and courts disaster.

Care should be taken to ensure that copper in concentrates does not exceed 15 parts per million.

Vitamin/mineral supplements should be added to feed in the proportions indicated; no more. It is illegal to add copper to sheep feed but it is inevitably present as a contaminant. If any doubt exists as to a feed's copper content, it must be analysed.

Avoid copper supplementation if animals are to be housed for any time. The copper availability from conserved forages and concentrate feeds considerably exceeds that from grass.

Take especial care with breeds known to be susceptible to copper poisoning. A number of expensive disasters have occurred, even without copper supplementation.

Figure 3.1   A lamb affected by daft lamb disease showing the characteristic star-gazing. (From Barlow, 1983)

Cripples

See Joint ill *and* Tick pyaemia.

Daft lamb disease

This problem is an inherited nervous disease which seems to be most common in the Border Leicester and the Scottish half-bred. The incidence in a flock is normally low.

Symptoms

This problem is normally evident soon after birth. In severe cases the lambs, which are in physically good condition, may be unable to stand.

In milder cases lambs may be able to stand and walk. The head is held high giving a 'star-gazing' appearance (Fig. 3.1). The lamb may walk in circles or apparently wander aimlessly. Affected lambs may not be able to suck from a ewe but will suck from a bottle. These lambs seldom thrive.

Treatment

There is no specific treatment. Mild cases can be reared with careful nursing, the nervous symptoms tending to regress with age.

Prevention

Affected lambs should not be retained for breeding. Consult your veterinary surgeon since this condition can be confused with swayback, Border disease and stiff lamb disease.

Diarrhoea

See Coccidiosis, Enteritis, Lamb dystentery *and* Parasitic gastroenteritis.

Enteritis

Enteritis means an inflammation of the lining of the gut. This results in a movement of fluid into the gut and increased gut movements. Scour (diarrhoea) is the result. Enteritis in lambs may be caused by lamb dysentery (see separate section), other infections including *E. coli*, rotavirus and *Cryptosporidium*, or by digestive upsets resulting from a sudden change of diet, such as when a lamb which is being fed milk replacer is fostered onto a newly lambed ewe. The remainder of this entry is devoted to enteritis not associated with lamb dysentery.

Symptoms

Digestive enteritis

Scouring is the obvious symptom but the lamb generally appears bright.

E. coli infection

K99-positive *E. coli* cause acute diarrhoea in newborn lambs as young as two days. Affected lambs have a profuse scour and are clearly ill.

Rotavirus infection
Diarrhoea is the main feature and, provided dehydration is prevented, lambs remain fairly bright.

Cryptosporidium infection
This infection can occur alone or along with other infections. It often occurs in lambs 7–10 days old. With this infection, lambs are distinctly ill, stop sucking and assume a tucked-up appearance.

Clinical signs give some clue as to the cause of enteritis but a precise diagnosis depends on laboratory examination of faeces or a post-mortem examination. A precise diagnosis has implications for treatment and prevention in the future.

Treatment

The successful treatment of enteritis is nine-tenths nursing and only one-tenth drugs. It is most important that dehydration and starvation should be prevented (*see* Hypothermia). Affected lambs should be fed, three times daily, with a solution containing both glucose (for energy) and electrolytes (to replace salts lost in the scour) by stomach tube (p.159). Hypothermia is a common complication of enteritis and the lamb's temperature should be checked if there is any doubt (p.154). Since the cause of the enteritis may be infectious, scouring lambs should be isolated. Contaminated pens should not be used for more lambs. After treating affected lambs, the hands should be washed and equipment such as stomach tubes sterilised before further use.

Digestive enteritis
No other treatment should be required.

E. coli infection
Antibiotic treatment will be required. Preventative antibiotic treatment to in-contact lambs may well be indicated.

Rotavirus and *Cryptosoporidium* infection
No other treatment should be required. Your veterinary surgeon may advise antibiotic treatment to prevent secondary bacterial infection.

Prevention

Ensure that all lambs receive plenty of colostrum within a few hours of birth. Keep the bedding in lambing pens clean. Ideally, the bedding should be changed between ewes but at least ensure that fresh straw is added. In severe outbreaks, oral antibiotics can be used under veterinary supervision as a preventative measure, but this is a last resort and is unlikely to have a permanent beneficial effect. It should be remembered that some infections in lambs can also cause disease in man – young children and the elderly are especially at risk. Take sensible precautions – wash your hands. Bear in mind that enteritis is also caused by lamb dysentery. At the end of lambing, buildings must be thoroughly cleaned and disinfected. If lambing outside use a fresh site next year if possible. These precautions are desirable even if no enteritis problems have occurred.

E. coli infection

Ewes can be vaccinated against this infection, the protection being passed to the lambs in colostrum (p.140).

Entropion

This is a turning-in of the lower eyelid (colour plates 3 and 4). If left untreated, the constant irritation caused by the inturned eyelashes leads to ulceration of the cornea (the front layer of the eye) and blindness finally results. This condition appears to be inherited, being more common in some breeds than in others.

Symptoms

In the early stages, a close examination may be necessary to spot this condition but very soon it becomes obvious. The affected eye 'weeps' and will often be closed. If both eyes are affected the lamb may be practically blind and unable to find the teat and suck. Eventually the cornea becomes white and opaque.

Treatment

This is a most satisfying condition to treat and the earlier the problem is spotted the better. In many cases the inturned eyelid can be simply 'flipped' into the correct position by pulling down the skin below the eye with the fingers but NOT by interfering with the eye itself. If this is not successful some surgical interference is necessary. Contact your veterinary surgeon about this.

There are a number of techniques which may be used. In the technique shown in colour plate 4 a surgical clip (Michel clip) is inserted in the skin below the eyelid. This draws the eyelid into the correct position. The clip may fall out of its own accord but if it does not it should be removed after seven days. If the eye has become infected (this is common), an antibiotic eye ointment should be applied for a few days (p. 163). Your veterinary surgeon will advise on this. While the lamb is recovering ensure that it is well fed – supplement by stomach tube if necessary (p. 155). Under no circumstances should entropion be left untreated – this would be the height of cruelty.

Prevention

Since this condition is thought to be inherited, affected lambs should not be retained for breeding. If the incidence is high, changing the rams should be considered.

Erysipelothrix infection

Erysipelothrix is the bacteria which causes swine erysipelas and erysipeloid in man. It is not a selective bacterium and also infects sheep.

The infection gains entry to the newborn via the navel and perhaps the gut, and in the older lamb via wounds and abrasions. The bacteria settle in the joints causing a painful arthritis. More than one joint is commonly affected.

Symptoms

Lame, stiff lambs are the normal sign. The joints may not be swollen and this condition is often missed in the early stages.

For a positive diagnosis, your veterinary surgeon must take samples for bacteriological examination.

Treatment

This problem responds well to antibiotic treatment provided that this is instituted at the earliest stage. Treatment for 7–10 days is advised to prevent recurrence. Treatment commenced later in the condition cannot repair damaged joints.

Prevention

Maintain good hygiene at lambing with plentiful clean bedding. Navels should be promptly dressed (p.188). Ensure lambs get early and adequate colostrum.

Take care to avoid dirty surroundings and instruments at operations such as castration, docking and shearing. In problem situations, ewes can be vaccinated against *Erysipelothrix*. If problems are experienced in older lambs, they can be vaccinated from six to eight weeks of age.

Exposure

See Hypothermia.

Eye infections

Infection of the eye can be a considerable nuisance in intensive lambing situations, the infection easily passing from lamb to lamb. If not promptly treated, infection can lead to temporary blindness and starvation.

Symptoms

In the initial stages, a discharge is seen – excessive tears. Soon a pronounced inflammation of the conjunctiva (the fleshy surrounding of the eye) develops (colour plate 5) and the eye may close. Finally, the cornea may become involved and become opaque. In a few cases the cornea may become ulcerated.

Treatment

Consult your veterinary surgeon for the correct antibiotic treatment for this problem (p. 163).

Prevention

The condition spreads from lamb to lamb by direct contact or via the surroundings, e.g. troughs.

Infected lambs should be isolated. In severe outbreaks the routine treatment of all lambs may be needed but the problem may reappear within a few weeks.

Eyelid (turned in)

See Entropion.

Faecal spoiling

This is a problem which all shepherds recognise. Quite simply the sticky dung of the lamb becomes stuck to the wool surrounding the anus and often the tail. In some cases the anal opening becomes practically blocked. Decomposing faeces next to the skin cause infection, inflammation and a thoroughly miserable lamb.

Treatment

In early cases the offending dung can be safely pulled off. In more advanced cases this will result in skin damage, and the faecal mass should be softened before removal by immersing the rear end of the lamb in a bucket of warm soapy water.

Prevention

Some shepherds associate sticky dung with either the type of ewe feeding or the use of milk replacers for lamb feeding – they may be right. This problem is easily reduced to no more than a nuisance by spotting affected lambs in the early stages.

Foot and mouth disease

The United Kingdom is currently free of this viral disease and hopefully will remain so. However, this freedom depends on the vigilance of all concerned with stock: the slightest suspicion must be reported immediately.

Foot and mouth disease is a highly contagious disease, it literally flies from farm to farm, either on the wind or carried by insects or birds. Thus even a few days delay in reporting a case can lead to extensive spread.

Symptoms

Lameness is normally the first sign seen. On a close examination excessive salivation and a discharge from the nose may be apparent. Vesicles (small raised fluid-filled swellings) may be found in the mouth and between the digits.

Treatment and prevention

Report your suspicions immediately. Delay is inexcusable.

Fractures

Bone fractures in newborn lambs are becoming increasingly common with more intensive lambing. The most commonly seen include fractures of the lower jaw as a result of careless interference during lambing, fractured ribs resulting from either crushing during lambing or accidents after lambing (most often being lain on by the ewe) and fractures of the legs, normally the result of careless handling.

Symptoms

These depend on the site of the fracture. A lamb with a fractured jaw will be unable to suck. Fractured ribs make breathing, and indeed any movement, painful. A leg fracture will, of course, result in lameness. This is most obvious in the case of a forelimb fracture, for the front legs support two-thirds of the lamb's weight. It may not be so obvious in the case of a hind-leg fracture where confusion with conditions such as joint ill may occur unless a careful examination is performed.

Treatment

The treatment of fractures is a complex subject depending not only on the bone or bones involved but also on the type of fracture present. Consult your veterinary surgeon. The correction of a jaw fracture is a difficult job and your veterinary surgeon may advise humane destruction. Fractures of the ribs nearly always necessitate humane destruction. Fractures of the legs can normally be treated by the application of some form of external support. In general, fractures of the lower leg are easier to treat than those higher up (Fig. 3.2) Your

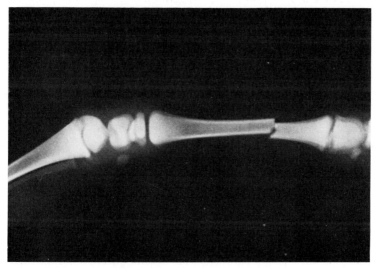

Figure 3.2   An X-ray of a fractured foreleg (metacarpus)

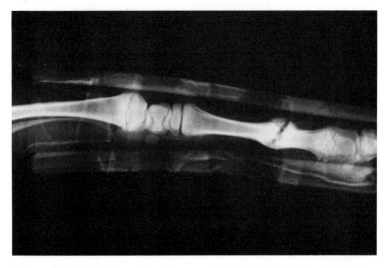

Figure 3.3 An X-ray of the fracture shown in Fig. 3.2 three weeks after treatment with Zimmer splints

veterinary surgeon may use either a plaster cast or Zimmer splints (Fig. 3.3). Zimmer splints are strips of aluminium lined with foam padding which are strapped to the lamb's leg. The lamb shown in Fig. 3.2 was treated in this way (Fig. 3.4).

Do not attempt treatment of a fracture. You may cause unnecessary pain and distress to the lamb.

Hairy shaker

See Border disease.

Hernia

See Umbilical hernia.

Husk

See Lungworm.

Hyperthermia

Hyperthermia means an abnormally high body temperature – the opposite of hypothermia. In temperate climates, such as prevails in the

Figure 3.4   The lamb referred to in Figs 3.2 and 3.3 two weeks after application of Zimmer splints

United Kingdom, this problem is only likely to arise when hypothermic lambs are rewarmed without adequate care and attention (*see* Hypothermia). This condition is rapidly fatal. A mild degree of hyperthermia may be associated with an infection (fever) but in this situation the lamb's temperature is steady and not rising rapidly, as is the case when a lamb is warmed excessively.

Symptoms

An affected lamb will appear weak and will pant in a fashion similar to that seen in the dog. This state quickly progresses to coma and death. A tentative diagnosis is easily confirmed by taking the lamb's temperature (p. 154). The normal temperature is 39–40°C (102–104°F). If the temperature of a lamb in a warming box is more than 41°C (106°F), it is hyperthermic.

Figure 3.5   A hypothermic lamb aged two hours. The hypothermia was caused by exposure

Treatment

Remove the lamb from the heat source, e.g. warming box. In mild cases this will suffice but in severe cases the lamb should be cooled with cold water. Take care not to 'overshoot' and cause hypothermia. Practically, this means stopping 'cooling' 1–2°C before normal temperature is reached.

Prevention

Common sense. The temperature of the warmer should be carefully regulated (never more than 37°C, 99°F) and the lamb's temperature should be checked while it is being warmed (p.170).

Figure 3.6   A hypothermic lamb aged 24 hours. The hypothermia was caused by starvation

Hypothermia

Hypothermia means a below-normal body temperature (normal for a lamb is 39–40°C, 102–104°F). This problem accounts for almost one-half of all postnatal losses. There are two distinct causes. The first is a high rate of heat loss from the wet newborn lamb aged up to about five hours – hypothermia due to exposure (Fig. 3.5). The second is a low rate of heat production in lambs aged more than six hours (and more commonly 12–72 hours) related to starvation and exhaustion of the lamb's body energy reserves (Fig. 3.6). Hypothermia due to exposure is more likely to occur outdoors, especially in bad weather, but it does occur inside, especially in the small, weak lamb, e.g. triplets or quads. Hypothermia due to starvation occurs both inside and out. All lambs

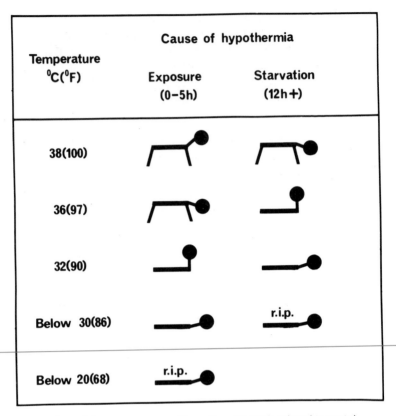

Figure 3.7   The appearance of hypothermic lamb related to rectal temperature and the cause of hypothermia

are susceptible to hypothermia of both types but in general the problem is much more common in twins and triplets and in lambs out of ewes in poor condition.

Lambs affected by hypothermia due to starvation have two problems. The first is hypothermia and the second is hypoglycaemia – a low level of blood glucose (sugar). This low blood glucose level must be corrected before the lamb is warmed. If it is not, it is likely that the lamb will die during warming; fit-like behaviour (often confused with recovery) is quickly followed by death.

Symptoms

The appearance and behaviour of the hypothermic lamb are related to both the cause of hypothermia and to body temperature (Fig. 3.7).

Lambs suffering from hypothermia caused by starvation tend to be weaker than those suffering hypothermia caused by exposure. This is caused by the low blood glucose level in the starving lamb. Diagnosis of hypothermia is a simple matter, provided a thermometer is used (p. 154). It is an expensive folly to rely on sticking one's finger (even an educated one) in the lamb's mouth.

Treatment

Treatment depends on both age and body temperature (Table 3.1). Full details of the techniques used will be found in Chapter 7, pp. 151–176, especially pp. 170–175.

Table 3.1 The treatment of hypothermia

Temperature	Age	Treatment
37–39°C (99–102°F)	Any age	Dry the lamb. Feed by stomach tube. Give shelter with ewe and other lambs. Check temperature again soon.
Below 37°C (99°F)	0–5 hours	Dry the lamb. Warm lamb in a warmer until temperature recovers to 37°C. Feed by stomach tube. Return to the ewe or transfer to 'weak lamb unit'.
Below 37°C (99°F)	More than 5 hours and able to hold up its head	Dry the lamb. Feed by stomach tube. Warm lamb in a warmer until temperature returns to 37°C. Feed by stomach tube. Return to the ewe or transfer to 'weak lamb unit'.
Below 37°C (99°F)	More than 5 hours and not able to hold up its head	Dry the lamb. Give intraperitoneal injection of glucose. Warm lamb in a warmer until temperature reaches 37°C. Feed by stomach tube. Return to the ewe or transfer to 'weak lamb unit'.

Prevention

This is largely a matter of common sense. Of prime importance is ewe nutrition. Good nutrition during pregnancy should ensure strong lambs with plentiful energy reserves and a ewe with plenty of milk. When lambing outside, the provision of simple shelter will reduce the risk of hypothermia due to exposure, and when lambing inside, especially at high lambing percentages, special attention should be paid to twins and triplets which are most likely to starve. Prompt use of the stomach tube (p. 159) will prevent many problems.

Inhalation pneumonia

In the newborn lamb the most common type of pneumonia is inhalation pneumonia, normally caused by careless bottle feeding. Weak lambs are unable to suck properly and when fed with a bottle a few drops of milk can easily enter the trachea (windpipe) and set up an infection in the lungs.

Symptoms

Commonly the lamb is found dead. If still alive the lamb is weak and breathing appears laboured.

Treatment

Vigorous antibiotic therapy is required – consult your veterinary surgeon. Treatment is, however, unlikely to be successful.

Prevention

Do not bottle-feed weak lambs – use a stomach tube (p.159). Careless bottle feeding is a dead loss. Note that pneumonia in lambs aged more than seven days is likely to be infectious in origin (p.76). A full veterinary investigation is required in these cases.

Iodine deficiency

Iodine deficiency is uncommon in the United Kingdom but it does occur in some areas such as Derbyshire. Iodine deficiency can also be induced by feeding 'goitrogenic' crops such as kale, rape and cabbage, which impair the use of iodine in the body.

Iodine is a major component of the thyroid hormones produced by the thyroid gland in the neck. These hormones are important in practically all body functions. Not surprisingly the most dramatic effects of iodine deficiency are seen in the foetus and newborn.

Symptoms

Signs include late abortion, the birth of weak lambs and the birth of lambs with enlarged thyroid glands, seen as a swelling in the neck.

Iodine deficiency rarely shows as clinical disease in adult stock, but it would seem likely that it would result in sub-optimal performance.

A diagnosis of iodine deficiency can be confirmed by examination of tissue at post-mortem examination or by analysis of blood samples.

Treatment

Oral supplementation is the normal route for supplying iodine.

Prevention

Iodine is routinely incorporated in concentrate feeds and in vitamin/mineral supplements. Care should be taken in feeding diets composed mainly or solely of brassica crops, especially in late pregnancy.

Jaw defects

Two defects affecting the lower jaw are found in newborn lambs: the lower jaw may be too short – undershot (Fig. 3.8); or too long – overshot (Fig. 3.9). The end result is the same. The lamb has difficulty in sucking and may starve.

Treatment

Clearly there is no specific treatment for these problems. Care must be taken to ensure that affected lambs are sucking and if they cannot they must be fed by stomach tube (p.159). After a day or so some lambs get the 'knack' of sucking from the ewe but others are unable to do this and must be artificially reared.

Figure 3.8   A lamb with an undershot lower jaw

Figure 3.9   A lamb with an overshot lower jaw

Prevention

Jaw defects are the result of some disturbance in foetal development. It is not known whether this is genetically controlled or not. If more than the occasional lamb is affected, the rams should be examined. Needless to say affected lambs should not be retained for breeding – at the very least they will find feeding difficult.

Joint defects

Occasionally defects of the lower joints of the legs, especially the forelegs, are encountered. The lamb may be unable to straighten the lower limb and 'knuckles' over (Fig. 3.10).

Treatment

In some cases application of a well padded splint or a bandage may help to extend the affected joints, but in very severe cases this is unlikely to be successful and humane destruction may be the most sensible course of action. Lambs with joint defects may have feeding problems and supplementation may be necessary (p. 155).

Prevention

Joint defects reflect an error of foetal development. Providing only the occasional lamb is affected there should be no cause for alarm.

Figure 3.10   A lamb with joint defects. The lower joints of the forelimbs
have not extended (straightened) properly

Joint ill

Joint ill is a bacterial infection of one or more joints which causes swelling,
pain and lameness. Unless treated promptly, permanent joint damage
results. The bacteria gain access to the lamb either via the navel at birth,
via wounds such as docking or castration, or via the gut especially when
colostrum is deficient or delayed. In tick-infested areas tick pyaemia can
cause joint ill in lambs aged over two weeks (*see* Tick pyaemia).

Symptoms

Joint ill is characterised by lameness, loss of appetite and general
depression. A close examination reveals pain and swelling in one or
more joints (Fig. 3.11).

Treatment

Prolonged antibiotic therapy is required.

Prevention

Ensure that lambs get plenty of colostrum. This will enhance their
resistance to infection. Navels should be dressed immediately after
birth to prevent the entry of bacteria by this route (p.188). Since the
lambs become infected with bacteria from their immediate
environment, a high standard of hygiene in the lambing house will
reduce the incidence of this problem. Instruments used for docking and
open castration should be kept clean.

Figure 3.11   A lamb with joint ill

Lamb dysentery

Lamb dysentery is a fatal disease affecting lambs in the first two weeks of life. The disease is caused by a bacterium called *Clostridium perfringens* Type B which multiplies in the gut and releases toxins (poisons) which cause the lamb's death.

Symptoms

Lambs may simply be found dead. More often affected lambs appear dull, do not suck and develop a blood-stained scour. Death follows within a few hours.

Treatment

There is no effective treatment.

Prevention

Fortunately this horrific disease can be prevented by vaccination of the ewe. Protective antibodies pass to the lamb in the colostrum. If either the ewe has not been vaccinated or the lamb does not receive colostrum, the lamb should be injected with antiserum soon after birth. This disease should be part of history, but cases still occur. These cases are commonly associated with either non-vaccination of the ewe, faulty vaccination (wrong time or too low a dose) or failure of the lamb to suck adequate colostrum.

Legs (broken)

See Fractures.

Listeriosis

Listeria bacteria can cause abortion in ewes (p.124), encephalitis in ewes (circling disease) and septicaemia in lambs, normally less than three months old.

Symptoms

Infected lambs are often found dead. If noted earlier they appear dull and have a high temperature.

Treatment

Early cases will respond to antibiotic treatment. Prophylactic treatment of other lambs may be appropriate.

Prevention

Silage is a common source of infection for sheep.

Attention should be paid to silage quality and handling. Ensure that pH value is below 5 (acid), and ash content below 70 parts per million (minimal soil contamination).

With clamp silage the surface layers are the most likely to be contaminated, and if possible they should not be fed to sheep. Cattle are less susceptible to listeriosis than sheep and it is often suggested that spoilt or dubious silage should be fed to them. This must still entail a small risk.

Do not allow uneaten silage to lie in feeding passages. Exposure to the air will allow the pH value to rise and facilitate multiplication of *Listeria* bacteria.

Big bale silage from damaged bags should not be fed to sheep.

Liver abscess (Liver necrosis)

This disease affects lambs aged three days or more. It is caused by the entry of bacteria via the navel soon after birth and possibly bacteria entering via the gut, especially in lambs which receive either late or insufficient colostrum. The blood vessels from the navel run through the liver in which the invading bacteria multiply and cause abscesses (colour plate 6).

Symptoms

Affected lambs first appear a little dull but their condition quickly worsens and death follows within three days. Antibiotic treatment often provides an apparent 'cure' but once treatment is stopped a relapse is common.

Treatment
Prolonged antibiotic therapy is required.

Prevention
The navel should be dressed as soon as possible after birth (p.188) and lambing pens kept clean. Ensure that lambs get plenty of colostrum – this increases resistance to diseases such as liver abscess.

Lockjaw
See Tetanus.

Louping ill
Louping ill is a viral encephalitis (brain infection) of sheep and other animals. It is spread by the sheep tick. The disease is most serious when it occurs simultaneously with tick-borne fever (p. 90). Tick-borne fever depresses the lamb's immune system, allowing the louping ill virus to multiply unhindered.

Symptoms
Affected lambs may simply be found dead. In less acute cases symptoms progress from slight loss of balance to coma and death in 1 –2 days. In the few recovered animals some nervous signs may persist for months.

In lambs also infected with tick-borne fever other signs such as scouring may be seen. Mortality in these lambs approaches 100 per cent.

A precise diagnosis depends on post-mortem examination.

Treatment
Little can be done except careful nursing. Depending on the age of the lamb stomach tube feeding of either milk or a glucose/electrolyte fluid will be indicated (p.159). In some cases humane destruction will be required.

Prevention
Vaccination is the principal method of control. Lambs born to vaccinated ewes acquire protective antibodies in the colostrum which

will protect them certainly for their first spring. The vaccine is a single dose product normally given to lambs in the later summer to protect them from autumn infection (p. 141).

In endemic areas further vaccination may be unnecessary, natural infection serving to boost immunity. However, if the level of infection in an area is relatively low revaccination at two-year intervals may be indicated.

A second approach to preventing louping ill is to avoid infestation with ticks. If grazing on tick-infested land can be avoided in the spring and autumn risk periods, problems can be avoided but this is seldom possible.

An alternative approach is to use insecticide preparations. In the past, dipping has been used but this can pose mis-mothering problems where young lambs are present. The recent pour-on preparations present an easier method of using this approach.

The animals most at risk are bought-in tick-naive stock, both ewes and lambs. These animals must be vaccinated one month before they encounter ticks.

Lungworm

Lungworm infection with *Dictyocaulus filaria* is common in sheep in the United Kingdom, but seldom produces serious disease. The most probable reason for this is that the measures taken to control parasitic gastroenteritis also control lungworm.

The life cycle of the lungworm is similar to that of gutworms (see Parasitic gastroenteritis, p. 69) with some important differences. In the small intestine, ingested larvae penetrate the gut wall and make their way via the lymphatic system and the circulation to the lungs. They then locate in the trachea, the bronchi (main branches of the trachea) and the bronchioles (small bronchi). They then develop into mature worms measuring up to 10 cm long. The life cycle of the worm is completed when the adult female worm lays eggs which develop into immature larvae. These larvae are coughed up and swallowed, to pass out in the faeces.

Not surprisingly the presence of these worms causes irritation and coughing. The tissue damaged by the worms is highly susceptible to secondary infection with bacteria such as *Pasteurella*.

Symptoms

In mild cases the occasional cough is all that is seen. In more severe cases the coughing is accompanied by laboured breathing and weight loss.

Treatment
Most of the anthelmintics in common use against gutworms are effective against lungworms.

Prevention
Strategies advocated for the control of gutworms are effective against lungworm (p. 74).

Muscular dystrophy
See Stiff lamb disease.

Navel hernia
See Umbilical hernia.

Navel ill
Navel ill is a bacterial infection of the navel which may be restricted to this region but may also be associated with further infections such as joint ill, liver abscess or spinal abscess (see separate entries). A high incidence of this condition is commonly associated with bad hygiene.

Symptoms
Affected lambs appear 'off-colour' and a close examination reveals swelling and tenderness in the navel area. The removal of any scab may result in the release of pus.

Treatment
The navel area should be cleaned. If necessary clip away wool which has become encrusted with pus. Prolonged antibiotic therapy is required.

Prevention
Navels should be dressed immediately after birth (p. 188). Ensure lambs receive plenty of colostrum. Keep lambing pens clean and ensure fresh dry bedding at all times. In some lambing sheds the level of

bacterial contamination is so overwhelming that it is surprising that not all lambs become infected.

Neonatal ataxia

See Swayback.

Nephrosis

Nephrosis is a normally fatal degeneration of the kidney seen in lambs from 2 to 16 weeks of age. The cause is presently unknown, but simple bacterial infection has been ruled out. The only common factor between cases appears to be that only lambs at pasture are affected. Normally only 1 or 2 per cent of lambs are involved.

Symptoms

Symptoms vary somewhat with age. Young lambs become listless and stop sucking. This quickly progresses to a drunken appearance, followed by coma and death. Some young lambs scour. The total course of the disease is normally only a few days.

Older lambs commonly scour and although they can appear bright, they quickly weaken and die.

Treatment

There is no specific treatment. Supportive therapy should be given including fluids by stomach tube (p. 159).

Prevention

Nothing.

Open mouth

See Jaw defects.

Orf

Orf is a most distressing skin disease which affects both lambs and ewes. It is more correctly called contagious (spreading) pustular (pus-filled swelling in the skin) dermatitis (inflammation of the skin). Orf is caused by a virus and can affect lambs only a few days old. The problem can easily spread from sheep to man and from person to person. This is most important in the context of the family. An orf lesion on the shepherd's finger may be little more than a painful nuisance (colour plate 9): a lesion on a young girl's face may be a tragedy.

Symptoms

In lambs the disease is normally first seen as scabs on the lips (colour plate 8). Close examination reveals pustules at the corners of the mouth. The area affected increases and may become further inflamed by secondary bacterial infection. The scabs may interfere with sucking and starvation is a common end result. This situation can be further compounded if the infection spreads from the lamb to the ewe's teats, making suckling a painful process for her. Mastitis often develops. In some lambs a very serious type of orf may develop in which the infection spreads into the mouth and sometimes even further down the digestive tract.

Treatment

There is no effective treatment for orf. Antibiotic in spray or ointment form is used to control secondary bacterial infection. General nursing is most important, for starvation must be prevented. This condition is self-curing in 2–4 weeks. It is essential that you seek veterinary advice.

Prevention

Orf is a most contagious disease and immediate isolation of affected ewes and lambs may help to prevent its spread. Great attention must be paid to hygiene – pens and feeding equipment exposed to orf must be disinfected before next year's lambing. A live vaccine is available which helps to control this condition but vaccination of the ewe appears to confer little or no protection to the lamb. The use of this vaccine tends to perpetuate the condition in the flock, albeit at a low level. Take professional advice if this problem occurs.

Overshot jaw

See Jaw defects.

Figure 3.12 Adult *Trichostrongylus vitrinus* burrowing under the epithelium of the small intestine

Paralysis

See Joint ill, Spinal abscess *and* Swayback.

Parasitic gastroenteritis

Parasitic gastroenteritis is probably the greatest source of production loss in sheep farming. Chronic parasitism leads to poor growth rate in lambs, even when no clinical disease is evident. Despite the availability of numerous effective anthelmintic drugs parasitism is a common cause of death in lambs.

Roundworms and disease

Roundworms cause disease by physically attacking the lining of the gut. Much of this damage is caused by the emergence of developing larvae and burrowing of adult worms (Fig. 3.12) This damage reduces the absorption of nutrients, and also leads to protein loss into the gut.

Some worms, e.g. *Haemonchus*, cause further damage by sucking blood.

Heavy infestation leads to disease and in extreme cases death, due mostly to dehydration and anaemia.

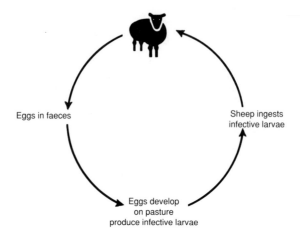

Figure 3.13   Basic parasitic roundworm life cycle

Life cycles

All roundworms, including lungworms, have basically the same life cycle (Fig. 3.13). No other animal is involved.

Development of eggs on the pasture is dependent on temperature and humidity. Ideal conditions are 18–26°C and 100 per cent relative humidity.

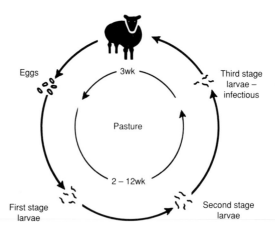

Figure 3.14   Life cycle of the common gut roundworms except *Nematodirus*

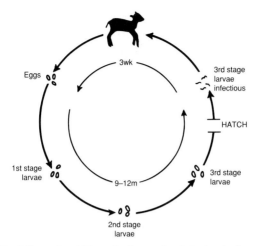

Figure 3.15 Life cycle of *Nematodirus,* a lamb-to-lamb disease

Infection persists over the winter as larvae and eggs on the pasture, and also as arrested larvae in the adult sheep gut. In the autumn larvae do not fully develop in the gut. Instead they encyst in the gut, only to emerge again in the spring.

The gutworms *Ostertagia, Trichostrongylus, Cooperia* and *Haemonchus* share a common life cycle (Fig. 3.14). Eggs in faeces hatch to yield free-living first stage larvae. These develop to third-stage larvae which are infective to sheep.

Nematodirus

Nematodirus, which can cause serious disease in lambs, has a slightly different life cycle (Fig. 3.15). Development of the infectious third-stage larvae depends on experience of a cold shock followed by warmth, i.e. winter followed by spring. To aid the survival of the larvae over the winter they remain within the eggs until 'hatching' in the spring.

The effect of this unique series of events is that infection in the spring depends on contamination of the pasture the previous year. The infection is practically a lamb-to-lamb affair, this year's lambs being infected by larvae derived from eggs shed by last year's lambs.

The nature of this life cycle has another important effect. As the temperature rises in the spring, all the 'hibernating' larvae on the pasture are subject to the same climatic influence and thus tend to hatch more or less simultaneously. Lambs are thus suddenly exposed to a massive *Nematodirus* infective load on the pasture.

In an early warm spring the 'hatch' as it is called may occur before the lambs are grazing. The larvae will die and there will be no disease. If the spring is cold the 'hatch' may be delayed until the lambs are older and have acquired some age-related immunity.

However, if the hatch occurs when young lambs have started to graze, i.e. about two months old, disease can be expected. The relationship of infection to climate enables the agricultural and veterinary advisory bodies to issue a *Nematodirus* warning when the risk is high. Inevitably this is a local warning, so take local advice.

Immunity to parasites

Adult sheep are largely immune to roundworm infection. They remain infected but at a very low level which has little or no deleterious effect. Adult infection with *Haemonchus* can sometimes seem to be an exception to this general immunity rule.

The immunity to parasites depends on previous infection and thus lambs have practically no immunity to parasitic gastroenteritis. They do however develop some age-related immunity and first infection, later in a lamb's life, is less likely to lead to disease than early infection.

This immunity in the adult can be compromised by two circumstances. First other sources of ill health, poor nutrition and other infectious disease will lower the immunity to gut parasites.

The second factor is lambing and lactation. From two weeks pre-lambing to six weeks afterwards, the ewe's immunity to parasites is depressed. This means that during this period the ewe can acquire more new infection and will pass more eggs out in her faeces, the so-called peri-parturient rise in faecal egg production.

An additional effect of the reduction in immunity is that larvae arrested in the gut wall the previous autumn become active. They develop into mature egg-laying worms.

This reduction in immunity causes no disease in the ewe herself, but it can lead to an increase in infection on the pasture, to which lambs are defenceless.

Sources of infection for lambs

There are two sources of infection for grazing lambs. First larvae which have over-wintered on the pasture from the previous year. These larvae become infective in the spring and early summer, but for practical purposes very few will remain after the end of May.

The second source of infection is the ewes with an increased egg output during the peri-parturient period just before and six weeks after lambing. This egg output will be derived from worms which have over-wintered in the ewe, and fresh infection acquired from the pasture in the spring.

This ewe-derived egg output onto the pasture does not pose an immediate threat to lambs. But the non-immune lambs act as multipliers. A relatively low level of infection in the lamb results in the production of large numbers of eggs in the lamb faeces.

This multiplication cycle continues until July/August when pasture contamination reaches danger levels. Lambs now meet a massive infective challenge. The result is parasitic gastroenteritis – diarrhoea and disease.

Control strategies against parasitic gastroenteritis must be aimed at preventing this dangerous multiplication of infection.

The exception to this sequence of events is *Nematodirus* (see above) where lambs are faced with a massive challenge in the spring. Adults play little part in the *Nematodirus* cycle except that they do remain infected at low levels. Thus when ewes are introduced to sheep-naive pasture the potential for *Nematodirus* disease in the future is present.

Symptoms

Heavy infection can lead to obvious disease with profuse scouring. Dehydration follows and this may lead to the death of the lamb. Sub-clinical infection which produces no obvious disease can have severe effects on growth rate, even if lambs are receiving regular anthelmintic treatment.

The blood-sucking worm *Haemonchus* infects the abomasum and rarely produces scouring. The loss of blood leads to anaemia with pale mucous membranes seen in the gums, conjunctivae and vulva. Lambs become weak and may develop a swelling under the jaw.

A diagnosis of parasitic gastroenteritis can be confirmed at post-mortem by counting the worms, and in the live lamb by a faecal egg count and blood samples. A faecal egg count is meaningless within three weeks of anthelmintic treatment. In young lambs especially, parasitic gastroenteritis is often confused with coccidiosis.

Treatment

Affected lambs must be treated with anthelmintic. Remember that other lambs in the group are likely to be similarly infected. After treatment, lambs should be moved to clean pasture (see below). If they must remain on heavily contaminated pasture further treatment will be required. Steps must be taken in following years to avoid this situation. Severely affected lambs will require fluid by mouth to combat the dehydration (p. 159),

Prevention

Prevention of parasitic gastroenteritis is highly desirable from both disease and production points of view. The aim of any preventative strategy must be to avoid exposing lambs to heavily infected pasture. Waiting for the lambs to scour is shutting the stable door after the horse has bolted.

Alternate grazing

The success of alternate or clean grazing in the control of parasitic gastroenteritis depends on two factors. First, very few larvae survive on the pasture for more than 12 months, and secondly sheep and cattle do not generally share common parasites. The pasture grazed by cattle in the previous year can be said to be safe for sheep, and vice versa.

There is one important exception to this rule – *Nematodirus*. Although *Nematodirus battus* does not usually cause disease in cattle it can survive in this host and cattle can thus perpetuate the infection on a pasture from year to year.

Many farmers cannot hope to practise an alternate or clean-grazing system but that does not mean that the concept inherent in this system need be abandoned.

Basically, any pasture that has been free of sheep for 12 months can be regarded as safe for young lambs (remembering the exception of *Nematodirus*). But other pasture can be added to this category depending on grazing history.

At one extreme pasture grazed by last year's undosed lactating ewes and undosed lambs is quite definitely not safe. But pasture grazed as aftermath by dry, dosed ewes can be regarded as safe.

If an alternate grazing system can be practised, lambs should be dosed in the spring to combat *Nematodirus*. Take advice on timing. It is usual to dose ewes at least once annually, preferably after lambing before movement to clean pasture.

In some systems more dosing may be required. For optimum results, good growth rate and no wastage of anthelmintic, the dosing regime must be tailored to the individual farm. There is no universal plan. Take professional advice.

On some lowland farms, a three-year grazing cycle can be employed using cropping as the third type of land use. Inevitably, such a system will be more effective than simply alternating cattle and sheep on a two-year system.

Control with anthelmintics

On farms which cannot practise alternate grazing and where sheep must graze the same permanent pasture, year in year out, parasitic gastroenteritis must be controlled by using anthelmintics.

The aim must be to prevent contamination of pasture reaching danger levels.

The peri-parturient ewe is the major source of initial pasture contamination. These ewes should be dosed at lambing (alternatively housed ewes can be dosed pre-housing), and then 3 and 6 weeks later. Lambs should be dosed at 6 weeks of age and then 3-weekly until the end of May. By this time most over-wintered larvae will have died out and the ewes will be producing very few eggs. If this type of dosing regime is repeated annually, general pasture contamination will be reduced and it may be possible to reduce the number of anthelmintic doses. This situation requires careful monitoring and you should consult your veterinary surgeon.

Use of anthelmintics

Anthelmintics are most valuable drugs and great care should be taken in their use.

Recently reports of anthelmintic resistance, i.e. worms not being killed by the normal therapeutic dose, have come to light, and this is cause for some concern. Most of the reports in this country have concerned the benzimidazole group of anthelmintics but abroad all classes of anthelmintics have been implicated.

Precautions should be taken to avoid this problem:

1. Ensure that all stock receive the full dose. In groups of lambs this means dosing for the heaviest lamb. Weigh a few lambs to make sure.
2. Do not dose unnecessarily.
3. Check dosing equipment to ensure that the full dose is given.
4. Alternate the class of anthelmintic annually. In the United Kingdom we have three classes of anthelmintic: the benzimidazoles, levamisole and morantel, and the ivermectins.

Alternate the groups on a three-year cycle. Changing from one member of a group to another of the same group will have no benefit.

In spite of (4) above it is advisable that all bought-in stock should be dosed with ivermectin. By this means benzimidazole-resistant worms (the commonest form of resistance in the United Kingdom) should be eliminated.

Pasteurellosis

In recent years *Pasteurella* pneumonia in lambs seems to have become more common. The responsible bacterium, *Pasteurella haemolytica*, is the same organism which causes pneumonia in adult sheep.

Symptoms

In lambs the disease is often very acute, the lamb being found dead. Signs in the sick lambs include dullness, a high fever, loss of appetite and laboured breathing.

Treatment

Prompt antibiotic treatment is indicated.

Prevention

Vaccines against *Pasteurella* pneumonia are available. Lambs from vaccinated ewes will acquire some protection in the colostrum. In problem situations the lambs may be vaccinated early in life. The first injection (two doses are required) can be given in the first week.

Pasteurellosis often seems to be precipitated by stressful situations e.g. transport, handling or bad weather. Lambs at pasture should be provided with shelter from inclement weather, and routine tasks should be undertaken in the least stressful way possible.

Photosensitization (Yellowses)

This problem is commonly seen on some Scottish hill farms, and arises when the skin becomes excessively sensitive to sunlight. This sensitivity is related to one of two mechanisms.

The first results from the ingestion of plants containing toxic substances which act directly on the skin, greatly increasing the absorption of sunlight energy. The second results from the ingestion of other plants which damage the liver. This liver damage results in a failure to break down phylloerythrin, a normal product of digestion in ruminants. This excess phylloerythrin acts on the skin to increase the absorption of sunlight energy.

Whatever the cause the result is tissue damage. Signs are seen on skin areas not protected by wool, notably the face and the ears.

Symptoms

Initially swelling is seen, often first on the ears. The affected skin dries and cracks. Dead skin eventually drops off, even the tip of the ear.

Treatment

There is no specific treatment other than housing, which must be used in serious cases. Application of creams or the use of injectable agents are unlikely to have any worthwhile effect.

Prevention

Several plants, including St John's wort, have been suggested as causes. If possible suspicious plants should be removed, but this is not feasible on extensive hill grazing.

Pine (cobalt deficiency)

Cobalt deficiency is a common cause of ill-thrift and poor growth rate in lambs. Cobalt is an essential part of vitamin B_{12}, a vitamin crucial to many mechanisms in the body. Vitamin B_{12} is made by the rumen bacteria and it is here that cobalt is required. Injection of cobalt into the animal has little or no beneficial effect.

Symptoms

Signs are non-specific. Poor growth rate leading sometimes to emaciation are the main signs. Lambs may show a watery discharge from the eye.

The problem of detection is similar to that seen with copper deficiency. Many animals are similarly affected.

Cobalt deficiency will have debilitating effects in adult animals but the effects are not so drastic. Poor productivity is likely.

Diagnosis depends on examination of blood samples taken by your veterinary surgeon. Examination for copper deficiency and chronic parasitic gastroenteritis will also be required.

Treatment

In acute cases vitamin B_{12} can be given by injection. This must be repeated monthly and other methods of supplementation are preferable.

Prevention

All concentrate feeds and vitamin/mineral supplements will contain cobalt, and animals in receipt of such feeds should not be at risk.

For grazing lambs cobalt can be applied directly to the pasture. Take advice on this measure as local conditions can affect the usefulness of this technique.

Cobalt (with selenium) is often incorporated into anthelmintics. If these agents are used monthly deficiency will be prevented, but in many situations (e.g. hill lambs), monthly dosing is neither practical nor desirable.

Cobalt can be given orally as a drench but monthly treatment is required.

Cobalt oxide pellets (bullets) are an established way of cobalt supplementation. The pellets should not be given to milk-fed lambs since an insoluble coating of calcium phosphate forms on the pellet. Care should be taken that the pellets are swallowed and not regurgitated (p. 146).

A further method of cobalt supplementation, especially useful where copper deficiency is also a problem, is the 'Cosecure' glass bolus. This can be given to lambs at two months of age (p. 148).

Poisoning

This is unusual in the newborn lamb since it obtains its nourishment from the ewe and is unlikely to eat poisonous plants. There are, however, two sources of poison of which the shepherd should be aware. The first is drug overdosage. Providing drugs are used as prescribed no problems should arise, but occasionally the understandable but erroneous logic that 'if a certain dose has a beneficial effect then double the dose must be better' is applied. This logic rarely applies and instead of getting increased benefit the deleterious effects of overdosage are seen. A second possible source of poisoning is phenolic disinfectants and dips. These are extremely toxic to newborn lambs and are very rapidly absorbed through the skin.

Older grazing lambs are exposed to all the perils facing adult stock. These include many poisonous plants including ragwort, yew, ivy, privet and rhododendron. Many incidents result from the careless disposal of domestic garden clippings 'over the fence'.

Most cases of chemical poisoning are man-induced. Copper and selenium toxicity are just two examples. Copper poisoning can result from excessive copper supplementation, high copper levels in concentrate feeds, the feeding of pig rations (which are high in copper) to sheep or even grazing pasture which has been treated with pig slurry.

Another common toxicity is monensin poisoning. Monensin can be used in-feed to control coccidiosis but excessive monensin is highly toxic.

Lead is no longer used in paint and this has reduced the incidence of lead poisoning. However in some areas (often old lead mining areas) high levels of lead can be found in soil and streams.

Symptoms

These are enormously variable depending on the poison, but nervous signs are common, either depression or excessive excitement followed by depression.

Treatment

In most poisoning cases no specific antidote is available (but note section on Copper poisoning). Treatment comprises nursing and the amelioration of symptoms by use of appropriate drug therapy. With plant poisoning it is sometimes appropriate to operate on animals which are known to have eaten the poisonous plant recently. A rumenotomy is performed and the rumen contents removed. This measure is only likely to be economic with valuable animals.

Prevention

In most cases this is common sense. Poisonous plants on extensive hill grazing may be difficult to cope with.

Premature birth

Premature birth is not in itself a disease, although it may be a result of a disease such as enzootic abortion. It may also result from poor nutrition or rough handling. It is a problem that every shepherd faces and so the principles of the treatment and care of these lambs are described here.

Figure 3.16 Premature triplets. Note poor birth coats and 'foetal' heads

Symptoms

Premature lambs are small, have poorly grown coats, are physically weak, may have teeth which have yet to erupt and often have 'foetal' heads (dome-like skull with narrow jaws) (Fig. 3.16). These features, plus the history of the ewe and flock (infectious abortion, poor nutrition), should make a diagnosis easy to make.

Treatment

The premature lamb is weak – it may be unable to suck or even stand. It has problems keeping warm and is a hypothermia risk. These lambs may have breathing problems, for the lungs sometimes fail to expand fully at birth. Premature lambs are also very susceptible to infectious disease. The premature lamb should be kept, if strong enough, with its ewe under cover in a clean sheltered pen. The lamb (not the ewe) should have access to an infra red lamp for extra warmth. If not sucking adequately, the lamb should be fed three times daily by stomach tube

(p. 159). If too weak to be left with the ewe the lamb should be kept in an individual box warmed by an infrared lamp (p. 172). Watch out for signs of other disease such as enteritis and treat promptly.

Prevention

This depends on the original cause of the premature birth. Prematurity is often a sign of a serious underlying problem – take professional advice.

Pulpy kidney

Pulpy kidney disease is not, as the name suggests, an infection of the kidney. It is a gut infection with clostridial bacteria. These bacteria produce a lethal toxin which is absorbed into the bloodstream. The name pulpy kidney derives from the post-mortem appearance of this organ.

Symptoms

Death is the commonest symptom. Live affected lambs show nervous signs such as increased excitability, progressing quickly to coma and death. Diarrhoea may be seen.

Prevention

The sheep clostridial vaccines are some of the safest, cheapest and most effective vaccines available. They must be used (p. 139).

Lambs from vaccinated ewes acquire antibodies in the colostrum which will protect them for the first 12 weeks of life. Vaccination of lambs can commence at eight weeks. If a lamb is known not to have consumed colostrum it can be passively protected with an injection of antiserum. This protection lasts 3–4 weeks. Thereafter the lamb must be protected by vaccination.

Pulpy kidney is associated with the consumption of excessive carbohydrate food. Little can be done to prevent lambs over-gorging themselves with milk, but care can be taken when introducing older lambs to concentrate feed. Introduce the feed slowly to give the ruminal bacteria a chance to adapt to the new diet. This procedure will also prevent ruminal acidosis (p. 84) and digestive upsets.

Rattle belly

See Watery mouth.

Redfoot

This is a distressing condition of the Scottish Blackface breed and its crosses, found most commonly in southern Scotland and northern England on heather hills. The cause is unknown but a genetic basis seems likely. Normally only 1–2 per cent of lambs are affected.

Symptoms

The horn on one or more hooves becomes detached revealing the sensitive tissues underneath. Secondary bacterial infection leads to severe lameness.

The condition may progress. The skin of the legs may become detached. The gums, mouth and lips, and even the eye may be affected.

Treatment

There is none. Affected lambs should be humanely destroyed.

Prevention

If the offending ram can be identified it should be culled. It would seem sensible not to keep future ewe-lambs from the dam of the affected lamb as flock replacements.

Redgut

This term describes the post-mortem appearance of young grazing lambs which have died suddenly. The cause of death is a twisted gut leading to blockage and bleeding.

The condition is associated with movement to lush pasture, which leads to excessive distension and movement of the intestines.

Symptoms

A dead lamb. If found alive affected lambs will appear very ill.

Treatment

In early cases surgical treatment could be attempted, but the chances of success are slim.

Prevention

Lambs and all sheep should be introduced to any change of diet, e.g. lush pasture, gradually. Either limit the time on the lush pasture or restrict intake by strip grazing.

Ribs (fractured)

See Fractures.

Rickets

Rickets is a member of the group of diseases known as the osteodystrophic diseases. In the lamb this term refers to abnormalities of the development of the skeleton.

Such abnormalities may be related to deficiencies of protein, calcium, phosphorus or vitamin D, or to a calcium:phosphorus imbalance. In addition, trace element deficiency such as copper deficiency can interfere with bone development (p. 42). In the United Kingdom, phosphorus deficiency and/or vitamin D deficiency are most commonly implicated.

Symptoms

Lameness and loss of condition are the common signs. The long leg bones, especially in the forelegs, may be bowed. Joints may be swollen.

Diagnosis depends on a careful examination of the lambs including a post-mortem examination. An X-ray examination may well be helpful.

Treatment

Once the severe signs noted above have been identified, a complete resolution of problems will take time. Affected animals should be given vitamin D by injection and attention paid to the diet to correct any deficiency. Only the correct dose of vitamin D should be given for vitamin D in excess is toxic. Parasitic gastroenteritis reduces the absorption of minerals from the gut, and must be treated at the same time (p. 69). If copper deficiency is present this also must be treated (p. 146).

Prevention

Vitamin D and phosphorus deficiencies are unlikely to occur in isolation. The occurrence of rickets in a flock suggests that total diet is inadequate in many respects. A total review is called for.

Poor conserved forage is often very low in vitamins, including vitamin D. Vitamin supplementation may be required if poor forage has to be used (p. 150). As noted under treatment, attention should be paid to preventing parasitic gastroenteritis and copper deficiency.

Roundworm infection

See Lungworm *and* Parasitic gastroenteritis.

Ruminal acidosis

Acidosis within the rumen, which is often fatal, is very definitely a man-made disease. The condition arises when grass-fed lambs are suddenly transferred to a cereal-based diet.

Rapid fermentation of the high carbohydrate diet leads to the production of excessive acid. Acid is absorbed into the body and water is drawn from the body into the rumen.

Symptoms

In mild cases animals look distinctly unwell and are unwilling to feed. They may scour. In severe cases lambs are found off their feet, breathing heavily and in very obvious distress.

Treatment

Severe cases are normally hopeless, and humane destruction is indicated. Mild cases can be treated by giving magnesium carbonate by mouth (up to 8 grams, four times daily). Supply plentiful clean water.

In valuable animals which are known to have recently ingested large amounts of cereal a rumenotomy can be performed and the cereal removed.

The concentrate food must be withdrawn from all the lambs at risk, and access given to good hay only.

Prevention

This is infinitely preferable to cure. Lambs should only be introduced to concentrate rations gradually with access to good forage. Beware of the lambs which don't eat the concentrate for a week, and then for no apparent reason gorge themselves. Remember that lambs are ruminants. This digestive system is designed to cope with relatively indigestible grass. Sudden exposure of this system to large quantities of highly digestible cereals is a recipe for disaster. How so many lambs survive some intensive finishing systems is something of a mystery.

Scad (Scald)

Scad is a disease of the feet caused by one of the bacteria found in foot rot. The disease is restricted to the cleft between the digits and no separation

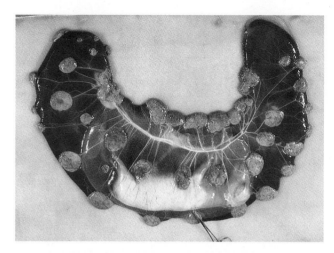

Plate 1
The contents of a uterus 90 days after conception. There is one foetus. The blood vessels running from the cotyledons to the umbilical (navel) cord can clearly be seen. (Picture by D. J. Mellor)

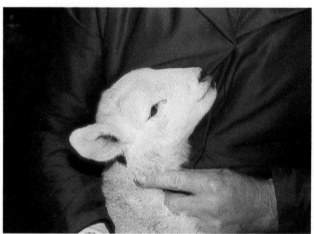

Plate 2
Entropion of the lower eyelid. The rim of the eyelid cannot be seen.

Plate 3
Entropion after treatment using Michel clips. The rim of the lower eyelid can be clearly seen.

Plate 4
Liver abscess.
(Picture by
K. A. Linklater)

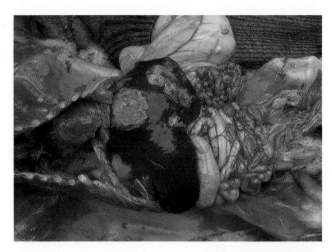

Plate 5
A lamb with an
umbilical hernia.

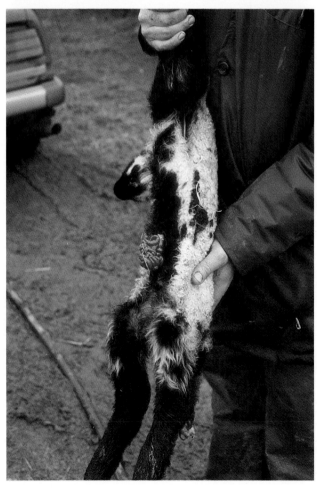

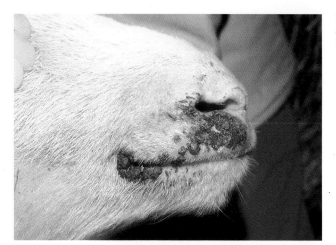

Plate 6
A lamb affected by orf.
(Picture by A. Inglis)

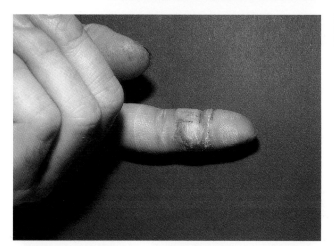

Plate 7
Orf on a shepherd's finger.

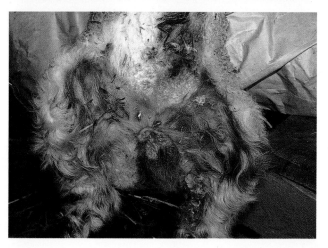

Plate 8
Castration with rubber ring, correct. The ring lies below the teats (highlighted in white).

Plate 9
Castration with rubber ring, incorrect. The ring lies above the teats (highlighted in white).

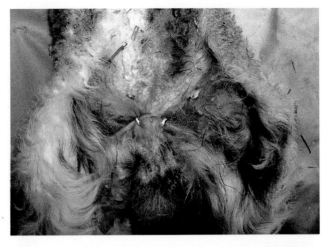

Plate 10
A lamb with tetanus. Note the stiff extended legs and the pricked ears. (Picture by M. J. Clarkson)

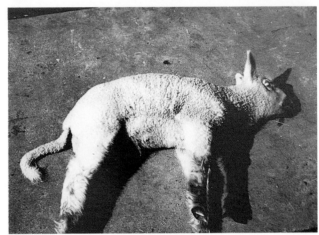

Plate 11
Prolapse of the uterus.

of the horn occurs. This problem in young lambs is nearly always associated with wet conditions underfoot. A change of environment is just as important as treatment if further cases are to be prevented.

Symptoms

Lameness, often very severe. Examination of the foot reveals an acute inflammation between the digits which is very painful to the touch.

Treatment

Topical application of antibiotic by spray is normally very effective. Keep the lamb back until the spray has had time to dry. Consult your veterinary surgeon for the best product to use.

Prevention

If inside, use more bedding. If outside try to find a drier pasture for the lambs. This will not only speed healing in affected lambs but will prevent the problem in other lambs.

Scour

See Coccidiosis, Enteritis, Lamb dysentery *and* Parasitic gastroenteritis.

Selenium deficiency

See Stiff lamb disease.

Sheepscab

This formerly notifiable disease of sheep, once eradicated, has returned as a constant on-going threat. Increased sheep movement and a failure to dip correctly must be two elements contributing to the failure of recent dipping policy.

Sheepscab is caused by the mite *Psoroptes ovis*. This mite spends its whole life cycle on the sheep, and spreads mostly from one sheep directly to another. The life cycle of the parasite can be completed in under three weeks, and with each mite laying up to 80 eggs it is easy to see how infection can take hold very quickly.

Symptoms

Intense irritation is the cardinal sign. Close examination of the skin reveals pustules (small fluid-filled swellings) which burst and 'weep' fluid. This fluid encrusts in the fleece. Fleece loss can be considerable.

Some infected sheep, which don't react as strongly to the mite, will show few or none of these signs. Thus an absence of symptoms does not necessarily mean an absence of infection.

A precise diagnosis is made by examination of skin scrapings when the mite is identified.

Treatment and prevention

Treatment is by means of a scab approved dip.

Prevention in the United Kingdom was by means of a compulsory annual dip. Whilst this measure did not eradicate the disease, it certainly contributed to its control.

Compulsory dipping has ceased and sheepscab is no longer notifiable. There is thus an increased onus on all flockmasters to watch for this dreadful disease.

An often forgotten benefit of annual dipping is control of other parasites such as lice and keds. These pests may well re-emerge on farms which cease dipping.

Slavers (Slavery mouth)

See Watery mouth.

Spinal abscess

This problem is seen in lambs aged from a few days to a few weeks. Spinal abscess has much in common with joint ill and liver abscess. Bacteria enter the lamb's circulation either through the navel or through the gut, especially in lambs which receive late or insufficient colostrum. In older lambs castration or docking wounds can serve as routes of entry. The bacteria lodge within the spinal column where an abscess forms. The abscess presses on the adjacent tissues and causes damage to the nerves and bones of the spine. A paralysis results. The exact symptoms seen depend on both the site and severity of the infection.

Symptoms

Most commonly the hind legs are affected. The lamb manages to move about on its forelegs, dragging the hind legs, but it quickly deteriorates.

Treatment
Prolonged antibiotic therapy combined with careful nursing is required. Treatment often produces a temporary remission of symptoms which return when the treatment stops.

Prevention
As with all such infections, a high standard of hygiene at lambing is imperative. Lambing pens should be kept clean and dry. Navels should be dressed at birth (p. 188). Ensure lambs get plenty of colostrum. Docking and castrating instruments should be kept clean. Spinal abscess can easily be confused with swayback. Take advice if in doubt.

Starvation
See Hypothermia.

Stiff lamb disease (Muscular dystrophy, White muscle disease)
This disease is caused by a deficiency of selenium and/or vitamin E. Lambs may be affected at any age up to six months but most commonly in the first month of life.

Symptoms
Lambs born to severely deficient ewes may be born dead or die suddenly in the first few days of life. Less badly affected lambs appear weak. Commonly the back legs become stiff and the lamb may eventually be unable to stand.

Treatment
Consult your veterinary surgeon who can prescribe an injectable preparation of selenium with vitamin E. It is most important to obtain an accurate diagnosis since this condition can easily be confused with other problems such as swayback and joint ill.

Prevention
If stiff lamb disease is confirmed treat all newborn lambs with selenium and vitamin E. In the future ensure that the ewes' diet is sufficient in

selenium and vitamin E. Your veterinary surgeon may advise treating the ewes with a preparation containing selenium and/or vitamin E. Note that selenium is toxic if given in excess.

Stones

See Urinary calculi.

Strike

See Blowfly strike.

Swayback

Swayback, sometimes known as enzootic ataxia or neonatal ataxia, is caused by a low availability of copper in the ewe's diet which inhibits the development of the lamb's nervous system during pregnancy. This low availability may be related to an absolute deficiency of copper in the diet or to an excess of the element molybdenum. Most cases of swayback occur in areas where copper deficiency is known to be a problem or on land which has recently been improved by liming, since this procedure reduces the availability of copper to the ewe.

Symptoms

The disease takes two forms in lambs. The first form is 'congenital' swayback. In this form the lambs are affected at birth. In severe cases the lambs may be unable to rise (Fig. 3.17), while in mild cases they may be merely a little unsteady on their hind legs. The second form of swayback is 'delayed' swayback. As the name suggests, the symptoms are only seen some time after birth, normally between two and six weeks, but sometimes as late as 12 weeks. Diagnosis of swayback may sometimes be easy but in many cases it is not. Other conditions which must be considered include spinal abscess and stiff lamb disease. Since the prevention of more cases depends on an accurate diagnosis, suspect swayback lambs should be submitted for veterinary investigation.

Treatment

Treatment of severe 'congenital' swayback is normally hopeless, and the lamb should be humanely destroyed. The recognition of 'delayed'

Figure 3.17 A lamb affected by swayback. (Picture by R. M. Barlow)

swayback suggests copper deficiency in the lambs (p. 42). Treatment with copper supplements (p. 146) will halt the progress of the condition.

Prevention

The prevention of swayback both at the time of an outbreak and in future years depends on the administration of copper-containing compounds to the ewes and, where appropriate, to the lambs. The ewe may be treated either by injection or by oral dosing with a copper oxide capsule or a Cosecure bolus (p. 148). Copper is a very toxic substance for sheep (p. 143). Consult your veterinary surgeon for both an accurate assessment of the problem and for instruction on the safe use of copper compounds.

Tetanus (Lockjaw)

This disease is caused by a toxin produced by the bacterium *Clostridium tetani*. Infection normally occurs via a wound, e.g.

castration or docking. The use of clostridial vaccines has greatly helped to reduce the incidence of this disease.

Symptoms

These are first seen 3–10 days after infection. Initially, the lamb appears stiff, is unwilling to move and muscle tremors may be observed. After 12–24 hours the limbs, neck and jaw become very stiff (colour plate 10). Disturbance of the lamb promotes increased stiffness and muscular spasms. Convulsions, failure of the breathing muscles and death rapidly follow. Occasionally only a mild form of tetanus may occur, there being little progress beyond the initial stiffness stage.

Treatment

Treatment of all but the mildest cases is useless. Affected lambs should be humanely destroyed for this is a most painful disease. Mild cases can be treated with antibiotics and tetanus antiserum – consult your veterinary surgeon.

Prevention

Vaccination of the ewe with clostridial vaccine effectively prevents this problem, providing the lamb gets adequate colostrum. The few cases that do arise are probably due to either incorrect vaccination of the ewe or failure of the lamb to take colostrum. Take care to castrate and dock lambs correctly under clean conditions. If there is any doubt about a lamb's colostrum intake, give tetanus antiserum by injection.

Tick-borne fever

All ticks are infected with the organism of tick-borne fever, *Cytoecetes phagocytophila*, and all lambs in tick areas will meet the infection. The infection alone is of limited consequence, but when it occurs simultaneously with louping ill (p. 64) or tick pyaemia (p. 91) the effects can be devastating.

Tick-borne fever depresses the lamb's immune system allowing the other infections to proceed unhindered.

Symptoms

Little is normally seen. The lamb will suffer a high temperature and will be 'off colour' for a week or more. There may be a moderate weight loss.

Treatment

This is seldom attempted. A single injection of long-acting oxytetracycline is effective.

Prevention

Tick-borne fever cannot be prevented but the time of infection can be delayed. Hopefully such a delay will reduce the severity of the tick pyaemia which commonly follows tick-borne fever.

Two complementary measures can be employed: a treatment with insecticide, commonly a pour-on preparation, and a double-dose injection of long-acting oxytetracycline. These measures are employed before the lambs are exposed to tick pastures.

Tick infestation

In the United Kingdom the sheep tick is most important because of the diseases it carries, tick-borne fever and louping ill. However a heavy tick infestation will itself have deleterious effects, related to the blood-sucking activity of the parasite.

Symptoms

Affected lambs will appear weak. Loss of blood will lead to anaemia, seen as pale gums, conjunctiva or vulva. Ticks will be much in evidence.

Treatment

In heavy infestation dipping will provide the most immediate response. Supportive treatment may be required.

Prevention

Use of a pour-on insecticide should prevent this problem.

Tick pyaemia

Staphylococcus aureus is a commensal bacterium of the lamb's skin which normally does no harm. But if the bacteria gain entry to the lamb's body whilst immunity is depressed by tick-borne fever the results can be most serious.

Symptoms

These vary with the site within the body where the bacteria lodge. Joint ill (cripples) and spinal abscess are two possible sequelae.

Treatment

Intensive antibiotic treatment is required.
Regretfully many lambs will remain 'poor doers'.

Prevention

The main problem is tick-borne fever. See page 90 for details of delaying the onset of this problem.

Trembling

See Louping ill.

Umbilical hernia

There is a 'gap' in the muscles of the body wall in the navel region through which the blood vessels from the placenta gain access via the navel cord to the lamb's circulation. This 'gap' should close very soon after birth. Occasionally it does not, and it may be big enough to allow the 'loose' contents of the abdomen, the intestines, to find their way outside the lamb's body – an umbilical hernia. Once some intestines are out, others follow. The ewe often exacerbates the condition by persistently licking the herniated intestines.

Symptoms

An impending hernia may be seen as a small swelling at the navel but more commonly the problem is only first noticed when a few inches of intestine have already escaped (colour plate 7). Sometimes these may be hidden inside the membranes of the cord. At this stage the lamb itself shows no untoward symptoms.

Treatment

Do not attempt this yourself. It is normally impossible to return the herniated intestines through the small hole and any attempt will only make matters worse. Protect the intestines from damage and try to keep them clean. Loosely wrap the lamb's abdomen with a clean towel.

Take the lamb immediately to your veterinary surgeon. Unless this problem is quickly corrected the lamb's condition will deteriorate at an alarming rate. Death in under 12 hours can be expected in untreated cases. Under anaesthesia your veterinary surgeon can enlarge the 'gap' in the body wall, gently replace the herniated intestines and then close the wound with sutures.

Prevention

This problem may have a genetic component and it is probably unwise to keep treated lambs as replacements. Take professional advice on prevention if more than the occasional lamb is affected.

Undershot jaw

See Jaw defects.

Urinary calculi (Stones)

This is another condition common in intensively fed lambs. It is relatively unusual in grass-fed lambs. Only ram lambs are affected.

Precipitates initially form in the kidney, and make their way to the bladder and the urethra (the tube leading from the bladder to the tip of the penis). It is in the narrow male urethra that problems are caused. The small calculi, often called sand, block this tube preventing the passage of urine.

Symptoms

These are mostly related to a full bladder and fruitless attempts to empty it. The lamb is in obvious discomfort and strains with no result. The abdomen becomes distended as the bladder enlarges. Blood, drops of urine and crystals may be seen at the prepuce.

As time passes the symptoms become more acute and the lamb is clearly in great pain.

Treatment

Veterinary help is required. Either medical or surgical treatment can be attempted, but the prognosis is poor. In some cases the blockage is just behind the urethral process (the vermiform appendage) on the tip of the penis. Removal of the process can relieve the problem.

Prevention

Once more prevention is infinitely preferable to cure.

A number of factors have been identified which predispose to urinary calculi:

1. A high mineral content in the diet, especially phosphorus and magnesium.
2. A low volume of concentrated urine related to the low water content of concentrate feeds.
3. A high urine pH. Urinary calculi are less likely to form when the urine is acid, pH value less than 7.
4. Urinary calculi are more common in castrated than entire ram lambs.

With these points in mind a number of measures can be formulated to prevent urinary calculi:

1. If possible leave ram lambs entire.
2. Introduce lambs to the concentrate diet gradually and always allow roughage, e.g. hay. This will maintain water intake.
3. Add sodium chloride (common salt) to the diet to increase drinking (1 per cent).
4. Add ammonium chloride to the diet to produce an acid urine.
5. Do not allow access to extra minerals.
6. Ensure that the calcium:phosphorus ratio in the diet is at least 2:1.
7. Ensure that lambs are used to drinking water from a trough before they are introduced to concentrate feed. Clean water must always be available.

Vitamin E deficiency

See Stiff lamb disease.

Watery mouth (Rattle belly, Slavers, Slavery mouth)

Watery mouth is a disease of intensive husbandry found in lambs aged 12–72 hours. Up to 50 per cent of all lambs can be affected.

In years gone by, watery mouth has been attributed to a host of causes ranging from constipation to castration with rubber rings. We can now discount the mythology which surrounds this disease for the cause is known.

Watery mouth is a form of endotoxic shock. Endotoxic shock or endotoxaemia arises when a large number of bacteria, commonly *E.*

coli, die in the gut. One of the breakdown products of dead bacteria is a toxin–endotoxin. When this endotoxin is absorbed into the bloodstream the result is watery mouth.

How do young lambs acquire this heavy bacterial load? Three factors contribute to this:

1. In the inside lambing situation the lamb is born into a heavily contaminated environment. Both the bedding and the ewe's fleece are rich in bacteria such as *E. coli.* When the lamb first attempts to suck, it often starts with a mouthful of fleece, not a teat. Its first suck is bacteria, not colostrum.

2. Unlike the adult sheep, the inside of the lamb's stomach, the abomasum, is neutral, pH 7. This condition is desirable for it allows the antibodies in colostrum to pass undamaged into the small intestine, whence they are absorbed. But it also allows bacteria to pass unhindered.

3. Colostrum is highly effective at preventing watery mouth, presumably by preventing the rapid multiplication of bacteria. But many lambs, especially twins and triplets, do not receive adequate colostrum early in life. It is not surprising that watery mouth is considerably more common in multiple lambs than in single lambs, and also in lambs out of ewes in poor condition.

In the early stages of this condition the passage of food through the gut slows down and may stop totally. The lamb ceases to suck. Gas accumulates in the abomasum and the lamb may become bloated (Figs 3.18 and 3.19). If these lambs are gently shaken, a rattling or tinkling sound will be heard, hence the name 'rattle belly'. A moderate amount of gas in the stomach can be deceiving for it gives the lambs a 'full-of-milk' appearance whereas the lamb is in fact starving.

The combination of starvation and endotoxaemia quickly kill the lamb.

Symptoms

Initially the lamb looks miserable and 'tucked-up'. The characteristic 'watery mouth', which is simply a drooling of saliva, soon appears (Fig. 3.20). The lamb ceases to suck and may become bloated. If not treated the lamb quickly deteriorates and dies.

Treatment

This should be started at the earliest possible opportunity. Inject the lamb once daily with antibiotic. Feed the lamb three times daily, by

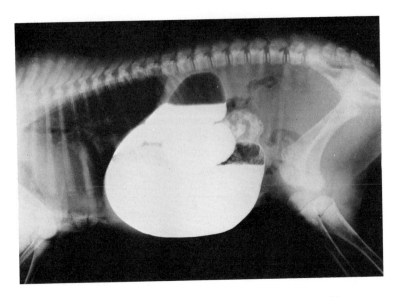

Figure 3.18 An X-ray of a healthy lamb given a 'barium meal' by stomach tube. The extent of the abomasum is indicated in the tracing below

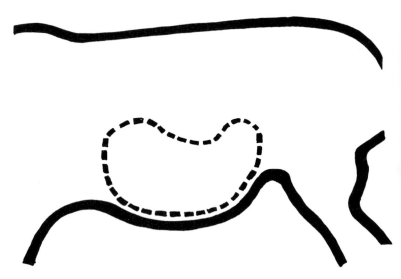

stomach tube, with 50 ml of a glucose/electrolyte solution (p. 159) containing oral antibiotic. If the lamb is not sucking from the ewe, increase the feed volume to 100–200 ml per feed. Continue treatment

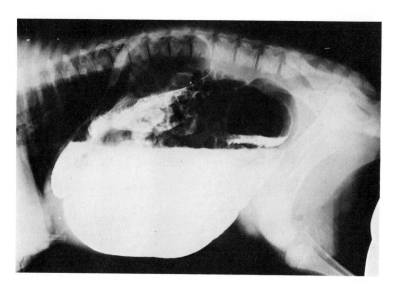

Figure 3.19 An X-ray of a lamb with watery mouth which has been given a 'barium meal'. The extent of the abomasum is indicated in the tracing below

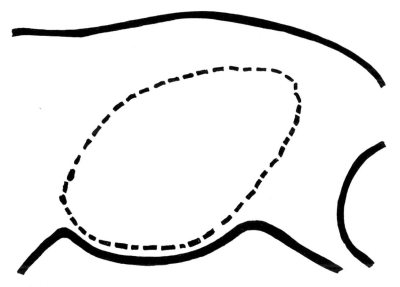

until the symptoms have gone. It is not advisable to feed watery mouth lambs with milk. Consult your veterinary surgeon about which antibiotics to use.

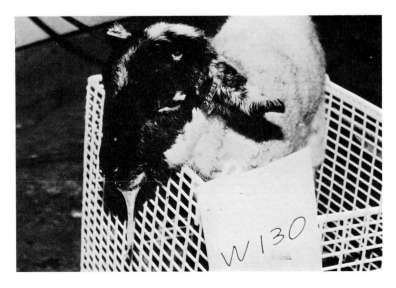

Figure 3.20 A lamb affected by watery mouth showing the characteristic excessive salivation

Prevention

Good ewe nutrition should result in a plentiful supply of colostrum. This will help to prevent watery mouth in addition to many other problems. Ensure that all lambs, and especially twins and triplets, take plenty of colostrum in the first hour of life. When necessary give supplementary colostrum by stomach tube. Keep lambing pens clean – this will help to prevent watery mouth and many other infections. Do not castrate lambs with rubber rings until 12 hours of age and preferably not until 24 hours, since the discomfort caused by this procedure reduces colostrum intake. When a serious problem does arise, lambs must be treated immediately after birth with antibiotic – treatment a few hours after birth may well be too late.

White muscle disease

See Stiff lamb disease.

Wounds

Most skin wounds in young lambs are of no serious consequence. Problems can arise through bacterial infection if the wound is not kept clean.

Treatment

If a wound is large, i.e. very deep or more than half an inch long, it may require stitching and you should consult your veterinary surgeon. Otherwise clip the wool surrounding the wound and bathe it with warm water containing a non-irritant disinfectant (follow the instructions on the bottle – using too strong a solution is harmful). Dry the area and apply a little antiseptic cream. If there is any doubt about the 'tetanus state' of the lamb (ewe vaccinated? lamb sucked plenty of colostrum?) give tetanus antiserum. Check the wound over the next few days to ensure that it is healing. If necessary bathe and dress again. Do not allow fluid seeping from the wound to become encrusted on surrounding wool – this will encourage bacterial infection.

Yellowses

See Photosensitization.

Chapter 4

Problems in ewes

In this chapter we have included notes on the problems common in lambing ewes. Problems not specifically associated with lambing time are not covered.

Before lambing you should consult your veterinary surgeon on all the problems described below. You should discuss appropriate forms of prevention and treatment, and also draw up guidelines as to when you can safely proceed yourself and when it will be prudent to summon professional help.

Abdominal hernia (ruptures)

This is an occasional problem which is most likely to occur in late pregnancy in the older ewe carrying twins or triplets. A weakness and splitting can develop in the muscles of the body wall either in the midline (ventral hernia), or at the side (flank hernia or 'fallen side').

Symptoms

In the case of ventral hernia the floor of the abdomen drops almost to the ground – only the skin is retaining the abdominal contents (Fig. 4.1). The ewe will walk with great difficulty, if at all. With a flank hernia, a swelling is seen to one side of the lower abdomen but the ewe is not normally so severely incapacitated (Fig. 4.2).

On very rare occasions, a ventral hernia can be confused with an excessive accumulation of fluid in the uterus – hydrops. This confusion is resolved by BRIEFLY turning the ewe on her back. In ventral hernia, the abdominal contents 'fall back into the ewe' and the break in the body wall can be found with the hand. In hydrops this does not happen.

Treatment

There is no specific treatment for either of these conditions. You must take advice from your veterinary surgeon. The ewe is unlikely to lamb

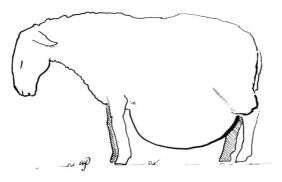

Figure 4.1   A ewe with a ventral hernia

naturally, especially in the case of a ventral hernia. If the ewe is near to lambing your veterinary surgeon may be able to induce her to lamb by injection of a hormone. Remember that the lambs will be premature. In severe cases where this is not possible, or when induction fails, the ewe may have to be destroyed on humane grounds.

Figure 4.2   A ewe with a flank hernia. The ewe is 'lop-sided'

Hypocalcaemia (Milk fever, Parturient paresis)

Hypocalcaemia means a low blood level of calcium. This condition is not an absolute deficiency of calcium, for the ewe has considerable reserves in her bones. It is rather a disturbance of the ewe's calcium metabolism in the last month of pregnancy or the first few weeks after lambing which results in a sudden and severe lowering of the blood calcium level. This is of serious consequence because calcium is essential for the efficient functioning of all muscles including the heart. The symptoms seen in this condition are largely those of muscular weakness. Before lambing, hypocalcaemia is normally associated with some stress such as being driven. Indeed stress alone, such as bad weather after shearing, can cause this problem at other times of the year. After lambing, the onset of lactation and the secretion of calcium into milk seem to precipitate this problem.

Symptoms

Initially just weakness may be seen but more usually the ewe is first noticed when she cannot stand. The ewe appears depressed and may rest her head on the ground. Occasional muscular spasms may be noticed. These symptoms may be confused with hypomagnesaemia or pregnancy toxaemia. However, the hyperexcitability of hypomagnes-aemia is missing and a prompt response to treatment with calcium solutions serves to differentiate hypocalcaemia from pregnancy toxaemia.

Treatment

Whenever hypocalcaemia is suspected a full dose of calcium solution should be injected under the skin (subcutaneous injection). A favourable response should be seen within an hour or so. Keep an eye on the ewe for the next few days. Relapse is not uncommon and a repeat treatment may be needed. If you do not see definite signs of recovery within an hour or so of treatment, you should question your diagnosis, and if in doubt take professional advice.

Prevention

The prevention of hypocalcaemia by dietary means seems attractive in principle, but there is little agreement between the experts on how it should be done. Some say give excess calcium so that there is plenty available to the ewe – others say restrict calcium intake to stimulate calcium release from the bones! What can be said with confidence is that both physical and nutritional stresses should be avoided before and after lambing.

Hypomagnesaemia (Grass staggers, Lactation tetany)

Hypomagnesaemia means a low blood level of magnesium. It is an acute condition most likely to occur in the first 4–6 weeks after lambing, although it can occur before lambing. It can be precipitated by moving the ewe to either lush or bare pasture and also by bad weather. Magnesium is essential for the function of nerves and most of the symptoms seen are nervous in origin. It must therefore be differentiated from other nervous diseases such as listeriosis, cerebro-cortical necrosis, and louping ill in tick areas. Hypomagnesaemia is commonly accompanied by hypocalcaemia and it is normal to treat both conditions if hypomagnesaemia is suspected.

Symptoms

Hypomagnesaemia is a much 'faster' disease than hypocalcaemia; the progression from apparent normality to death may only take an hour or so. In many situations the first thing noticed is a dead ewe. In the early stages the affected ewe appears excitable, may walk with a stiff gait and may show nervous twitchings or spasms. As the disease progresses the ewe goes off her legs and may lie with all four legs extended in spasm. Convulsions and death follow.

The problem can be confused with either hypocalcaemia or pregnancy toxaemia (before lambing only). Confusion with hypocalcaemia is of no immediate importance since both conditions are treated as a routine. Confusion with pregnancy toxaemia should be resolved by the fast progressive nature of the condition and hopefully prompt response to treatment.

Treatment

A calcium solution with added magnesium should be given subcutaneously. Your veterinary surgeon may also recommend the subcutaneous injection of a stronger solution of magnesium in addition. Under no circumstances should a strong magnesium solution be given by intravenous injection – this is a sure way of killing the ewe. Take advice from your veterinary surgeon about which solutions to use, when and how.

Prevention

If a sudden change of diet or pasture has precipitated cases of hypomagnesaemia, the careful reversal of the change is probably wise. In the longer term, the incidence of hypomagnesaemia can be reduced to a minimum by ensuring an adequate dietary intake, ideally in the form of a high magnesium concentrate. Provide shelter during the high risk period and ensure that ewes in poor condition receive extra rations.

In known high-risk situations hypomagnesaemia can be prevented by either pasture treatment or administration of a magnesium 'bullet' (p. 144).

Mastitis (Udderclap)

Mastitis means an inflammation of the udder, normally caused by bacterial infection. The form of mastitis which occurs in early lactation is an acute disease which can easily lead to the death of the ewe if not promptly detected and treated.

Symptoms

The ewe is first noticed either when she does not come to the feeding trough or when she limps on a hind leg as she tries to relieve the pain in her udder. An examination reveals that one side of the udder is swollen, hot and painful. The ewe is depressed and may well have a high temperature (more than 40.5°C, 105°F). The milk in the affected quarter is often thin and may contain clots of blood. Sometimes one's attention is first brought to this problem by hungry lambs, for mastitis severely depresses or stops milk production.

Treatment

Prolonged antibiotic therapy is required, and the sooner it is commenced the better are the chances of saving the ewe and the infected quarter (or should it be the udder half?).

Prevention

A number of factors are likely to increase the chances of a ewe getting mastitis. These include orf, bad hygiene in the lambing area and over-zealous sucking by the lambs. The first factor has been discussed earlier (Chapter 3) and the second requires no further comment. Over-zealous sucking, which can lead to teat damage and infection, is most likely to occur when the ewe has insufficient milk. The hungry lambs continue to suck the empty udder and as they become more frustrated they increase their efforts. In this situation the lambs should be supplemented until the milk supply increases, or removed and either fostered or artificially reared.

Metritis (Inflammation)

Metritis means inflammation of the uterus. This is usually caused by bacterial infection and is a serious condition which can kill the ewe. It occurs either when the ewe has aborted and the placenta (afterbirth) has been retained, when a dead rotten lamb has been born, or when the shepherd has assisted a lambing without observing all the precautions noted in Chapter 2.

Symptoms

The first sign seen is often a dull ewe. A close examination reveals a discharge from the vulva. This is often brown or green in colour and foul smelling. A high temperature (more than 40.5°C, 105°F) is a common finding.

Treatment

Prolonged antibiotic therapy is required.

Prevention

Abortion and the birth of rotten lambs are beyond the shepherd's immediate control, although they are causes for long-term concern. Consult your veterinary surgeon on the best preventative measures to take in these cases.

Some cases of metritis are caused by clostridial bacteria. Vaccination of the ewe with a clostridial vaccine should prevent these.

When aiding a ewe during lambing, be as clean and gentle as you possibly can. To prevent infections, antibiotic pessaries can be placed in the uterus once the lamb has been delivered, but these may later be expelled with the afterbirth. An injection of long-acting antibiotic is required.

Pneumonia

Pasteurella pneumonia can occur at any time of the year, but in some flocks it does occur more commonly around lambing. Sometimes a number of ewes are affected but often the occurrence is sporadic.

Symptoms

Ewes may be found dead. More commonly the ewe is found obviously ill with a high temperature and laboured breathing. Discharges may be seen from the nose and eyes.

Treatment

Antibiotic treatment is required: consult your veterinary surgeon. If only the odd case is seen, prophylactic treatment of other ewes is probably unjustified, but if an outbreak seems likely such treatment is worthwhile. Past flock history will be important in taking this decision.

Prevention

In flocks where *Pasteurella* pneumonia is a recurrent problem vaccination should be considered. Such vaccination can be conveniently given in preparations combining clostridial vaccines and pneumonia vaccine. The pneumonia vaccine, like clostridial vaccine, is a dead preparation and two doses are required. The last dose (or a booster dose) should be given at least four weeks before lambing is due (p. 143).

Outbreaks of *Pasteurella* pneumonia are often associated with stressful conditions and handling. Care should be taken to avoid such situations around lambing. Ewes should be moved to lambing areas well before lambing is due; moving them just before lambing often seems to precipitate problems.

Pregnancy toxaemia (Twin lamb disease)

Pregnancy toxaemia is a metabolic problem of the ewe which is found in the last four weeks of pregnancy – never after lambing. It most commonly occurs in the ewe carrying two or more lambs but it can occur in ewes, especially hill ewes, carrying only one lamb. This condition results from a shortage of energy. This energy shortage is related to the requirements of the growing lamb and to the nutritional state of the ewe. Not surprisingly, the condition occurs in ewes in poor condition on a low nutritional plane. But it also occurs in very fat ewes whose appetite may be depressed, and in ewes which are greedy trough feeders, but which take little hay or silage. Stress factors such as bad weather, handling and hard driving may bring on this problem in susceptible ewes.

Symptoms

In the early stages the ewe separates from the flock. Signs of blindness may be evident. The ewe soon becomes depressed, stops feeding and may show nervous signs. These are variable but can include 'head pressing', unusual carriage of the head, fine tremors, teeth grinding and even convulsions. Breathing may appear to be laboured. It may be possible to detect the sweet smell of acetone on the ewe's breath. After a day or so the ewe becomes recumbent (unable to rise). Regurgitated stomach contents may be seen in the nose and a scour may develop. The ewe may become blown. Coma and death follow.

Treatment

The chances of success are not high but they are best if treatment is started at the earliest possible stage. Assume that hypocalcaemia and hypomagnesaemia are also present and treat accordingly (pp.102–104). In the case of the overfat ewe, shearing may help – this stimulates the ewe to break down her own energy reserves. If the ewe is known to be near lambing, labour can be induced. Otherwise the principle of treatment is to give the ewe energy. Offer appetising food such as molasses and give propylene glycol by mouth. Your veterinary surgeon may supplement this regime by giving glucose by injection. In addition to feeding, a high standard of nursing is required. The ewe should be removed indoors to a deeply bedded pen. If she is recumbent move her at least twice daily to prevent the development of sores and pneumonia. An early improvement with treatment is a hopeful sign; further deterioration of the ewe's condition is not.

Prevention

The major principle involved is nutrition from before tupping until lambing. This requires proficiency in, and frequent use of, body condition scoring. Ewes should be in good, but not fat, condition at mid-pregnancy, i.e. a condition score of 3, and should receive an improving plane of nutrition as lambing approaches. Separate the lean ewes and give them extra rations. Shy feeders must be separated from the 'bullies'. This ensures that the shy ewes get enough feed and that the 'bullies' do not get too much. Ewes with either teeth or feet problems should be culled – if not, they require preferential treatment if their nutrition is not to suffer. When folding on turnips, feed concentrates first as some ewes are slow to eat concentrates after turnips. Check that hay or silage is of good quality and is palatable – a stomach full of rubbish is no good to a ewe in late pregnancy. If you are feeding more than 0.5 kg (1 lb) of concentrates daily, divide the ration into two feeds. Keep the stress of procedures such as driving and dosing to a minimum.

Prolapsed vagina

Prolapse of the vagina is a condition occurring in the last three weeks before lambing in which the vagina is pushed out through the lips of the vulva. It is most common in old fat ewes but can occur in ewes of any age. The prolapse can include the uretha (the tube connecting the bladder to the vagina) and also the bladder itself. This leads to an inability to urinate, pressure in the bladder and further straining which makes the condition worse. In severe cases the wall of the vagina may break and intestines may be herniated through the hole. These extreme cases are hopeless. Summon your veterinary surgeon who can painlessly destroy the ewe and maybe salvage the lambs.

Symptoms

The prolapse is normally first seen when the ewe is lying. It may not be present all the time – it can literally pop in and out. The condition may progress no further but it often does. Eventually the vagina becomes permanently prolapsed. The ewe strains, making the condition worse.

Treatment

A number of treatments have been advocated over the years. These include tying strands of wool across the vulva (impossible in many short-woolled breeds and in shorn ewes) and a number of patent restraining devices which in our experience often fall out (Fig 4.3). As long as the ewe can be closely watched for signs of lambing, the best method is to stitch the lips of the vulva. You should only attempt this if you are experienced and have received detailed instruction from your veterinary surgeon.

As noted above, a prolapse may block the uretha preventing urination. This increases straining. The urethal orifice is situated on the floor of the birth canal (i.e. nearest the ewe's feet) about two inches from the exterior. The full bladder may be profitably relieved by carefully elevating the prolapsed vagina with clean lubricated fingers above the floor of the birth canal for a few seconds. No force is required.

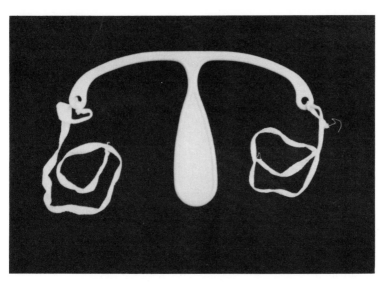

Figure 4.3   A retaining device used to treat a prolapsed vagina

Use nylon tape and a large sharp needle (often called a post-mortem needle) (Fig. 4.4). Correction of the prolapse and stitching are much easier if the rear end of the ewe can be raised, so reducing the pressure from the contents of the abdomen. The prolapse should be washed using warm water with a non-irritant disinfectant, and then gently pushed back using bent fingers. Three stitches are inserted (Fig. 4.5). The needle should not be passed through the lining of the vulva as this will cause straining – instead go in and out of the skin alongside the vulva. The ewe should be given antibiotic by injection and either her number noted or a permanent mark applied, for she should be culled after weaning. In severe cases it may be advisable to apply a truss in addition to the stitches. Once the ewe has started to lamb remove the sutures. If you don't, serious tearing will result.

More difficult cases will require veterinary attention. Your veterinary surgeon may give a spinal anaesthetic which will have the dual advantages of stopping straining and abolishing sensation.

Remember: DO NOT attempt this procedure unless you are very experienced.

Prevention

The cause or causes of this problem are not known. It has been attributed to sloping ground (unlikely), over-fatness, excessive bulky

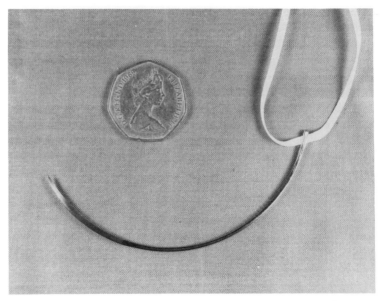

Figure 4.4   Needle and tape used in the treatment of a vaginal or uterine prolapse

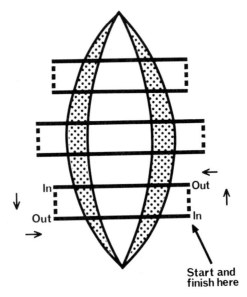

Figure 4.5   Stitching of the vulva after a prolapse (From *Management at Lambing*, 1983)

foods (turnips) and relaxation of the tissues in the vaginal area after many lambings. Prolapse is more likely to occur if the ewe is fed either poor quality roughage or excessive concentrates. Both these factors depress digestion and increase the pressure in the abdomen. Feed only good quality roughage and divide the daily concentrates ration into two feeds. Ewes in late pregnancy should always be handled gently.

Recent investigations have suggested that prolapse of the vagina is most common when nutritional control of the flock is poor.

A combination of regular monitoring of condition score, good roughage and adequate but not excessive concentrate feeding would seem the best approach to preventing this problem.

Prolapsed uterus

Prolapse of the uterus occurs immediately or soon after lambing. The whole uterus passes through the vagina and vulva and hangs from the ewe. The inverted uterus can be severely damaged if prompt action is not taken.

Symptoms

One look is normally enough, see colour plate 11.

Treatment

If you are not very experienced summon professional help immediately. A good job must be done first time. If a ewe prolapses for a second time the outlook is grim. Whilst waiting for your veterinary surgeon, keep the ewe still and wrap the uterus in a clean towel.

As with a vaginal prolapse, correction of uterine prolapse is much easier if the ewe's hindquarters are raised. Clean the surface, i.e. the inside of the uterus, with warm water containing a non-irritant disinfectant. Remove pieces of dirt. If the placenta is still attached to the wall of the uterus this can be removed if it comes away easily – if not, leave well alone. Lubricate the uterus with lambing lubricant and with the whole clenched fist push it back into the correct position. Suture the lips of the vulva as described for vaginal prolapse and apply a truss. The truss and sutures can be removed after ten days. Give antibiotic by injection and keep a close eye on the ewe.

We repeat: DO NOT attempt this procedure unless you have received full instruction from your veterinary surgeon.

Prevention

As for vaginal prolapse. It would seem wise to cull affected ewes after weaning.

Chapter 5

Abortion in ewes

In the last thirty years abortion in ewes, especially infectious abortion, has become increasingly important in the sheep industry of the United Kingdom. The annual cost probably exceeds £20 million. This cost is not divided equally between the nation's sheep farmers. Some have no problem; others have serious problems which can cost thousands of pounds in just one year. Table 5.1 lists the more important causes of abortion. The major infectious causes are enzootic abortion of ewes and toxoplasmosis (Table 5.2).

Table 5.1   Common causes of abortion in ewes

Infectious	*Non-infectious*
Enzootic abortion of ewes	Pregnancy toxaemia
Toxoplasmosis	Rough handling
Salmonellosis	Transport
Campylobacteriosis	Other stress
Listeriosis	
Border disease	

Non-infectious abortion

In ewes, 1 per cent but certainly not more than 2 per cent, may be expected to abort each year due to non-infectious causes such as stress of handling or transport, or pregnancy toxaemia.

However, even if a non-infectious cause is suspected, and even if the laboratory fails to demonstrate any infection, it must be assumed that such abortions are infectious. They must be treated as outlined below (p. 116). Only at the end of lambing will you be able to say that only 1 per cent of ewes aborted and that all the laboratory findings were negative.

Table 5.2   Abortion: a summary of the common infectious causes

	Enzootic abortion of ewes	Toxoplasmosis	Campylobacteriosis	Salmonellosis (not *S. abortus ovis*)	Listeriosis	Border disease
Cause	*Chlamydia* (small bacteria)	Protozoa	Bacteria	Bacteria	Bacteria	Virus
Species affected	Sheep (also man)	Many	Sheep and other animals	Many	Many	Sheep and cattle
Source of infection	Aborting ewes	Infected cat faeces	Aborting ewe, carrier ewe, rodents, birds?	Contaminated food or water. Cattle	Poor silage	Congenital. Carrier ewe
Abortion	Last 2 weeks Fresh. Alive or dead	Alive or dead. May be mummified	Last 7 weeks. Foetus may be swollen.	Last 6 weeks	Alive or dead	Alive or dead
Ewe effects at abortion	None	None	None	Varies with species None to death	May be metritis	None
Control of outbreak	Isolation. Antibiotic to pregnant ewes	None. Monensin in food may help in future	Isolation. Antibiotic to pregnant ewes.	Isolation. Antibiotic to pregnant ewes	Check silage	None
Immunity	Will not abort again. May retain infection	Life-long	Life-long	Good to species involved	Probably good	Good except congenital infection
Vaccination	See text	New vaccine	Farm vaccine can be made	Depends on species involved	None in UK	None
Culling policy	Few abortions – cull. Many abortions – keep.	Keep	Keep	Keep	Keep	Cull carrier ewes
Replacement policy	EAE-free stock	None	Graze with flock before tupping	None	None	Mix with ewes and lambs up to 2 months pre-tupping

Infectious abortion

Enzootic abortion of ewes

The cause

Enzootic abortion of ewes (EAE) is caused by a bacterial organism called *Chlamydia psittaci*. The organism principally infects the placenta, interferring with the passage of nutrients and oxygen from the ewe to the lamb. It also disturbs the production of placental hormones.

The disease

Infection can result in the premature (normally 10–14 days) birth of a fresh dead lamb, the premature birth of a weak live lamb or the near-term birth of a weak live lamb. Thus some cases of infection are not immediately apparent.

The principal source of infection is the aborting ewe. The foetus, placenta and uterine discharges are rich in infective organisms which may be ingested by clean ewes.

These freshly infected ewes do not usually abort. They harbour the infection without showing symptoms until the next lambing, when they in turn abort, serving as a source of infection for further clean ewes.

This is the normal sequence of events – infection one lambing, abortion the next. However, if a clean ewe is infected some time before lambing, say more than five weeks, she may abort in that pregnancy. Such a situation is only likely to arise if either lambing is unduly protracted, or if a farm has two flocks, one lambing early and one lambing late. Infectious abortion in the early flock could cause abortion in the late flock if the late flock ewes had access to the early lambing.

If a ewe has aborted once she is very unlikely to abort again. But future lambings (lamb, placenta and discharges) may be infected. She may still serve as source of infection for clean ewes.

A typical flock history of infection with EAE goes as follows:

Year 1:   clean flock with no history of EAE buys in infected replacement females.
Year 2:   some of replacement females abort, spreading infection to the rest of the flock.
Year 3:   an abortion storm, up to 30 per cent of ewes abort, all ages.
Year 4:   and subsequently: 5–10 per cent of ewes abort, mostly younger ewes.

The losses in Year 3 are catastrophic and this is followed by an annual loss which will make a considerable dent in flock profits. EAE can cause metritis in the ewe. Live lambs born to infected ewes or lambs fostered onto these ewes are likely to become infected, and should not be retained as replacements.

Action after an abortion

Action is based on the knowledge that the aborted ewe is a source of infection for other ewes – discharges, placenta and foetus or lamb. Although you may suspect EAE you cannot be sure that other infection such as salmonellosis is not present. You must assume that the aborted ewe is a danger for other ewes and people.

1. Isolate and mark aborted ewe.
2. Keep other ewes away from abortion site.
3. Seek veterinary assistance and send samples (lamb or foetus and placenta) to laboratory for examination. Until proved otherwise assume that the abortion is caused by EAE and salmonellosis.
4. Mark any live lambs from aborting ewes.
5. If further ewes abort treat as above, and send further samples to the laboratory. Do not assume that all abortions have the same cause.
6. If enzootic abortion of ewes is confirmed your veterinary surgeon may advise treating remaining pregnant ewes with long-acting oxytetracycline. This does not eliminate the infection but it can delay abortion, and thus increase the viability of lambs from infected ewes. Discuss the potential cost benefit of this treatment with your veterinary surgeon before proceeding.

Future flock policy

If only a few ewes have aborted and these ewes have been identified and isolated, there is very good argument for culling with the hope that the infection can be eliminated. However, if many ewes have aborted it is inevitable that the infection has been widely spread through the flock, and culling is probably pointless.

Until recently, a killed vaccine against EAE has been available. This product has been withdrawn. However, there would seem to be prospects of two vaccines in the future, an improved killed vaccine and a new live vaccine. Consult your veterinary surgeon on the current situation.

Management at lambing can help to reduce the spread of infection. The general direction of infection spread is from older ewes to younger ewes. It thus follows that if ewes are segregated on the basis of age during lambing, and if possible for three weeks after, the spread of infection will be greatly reduced.

Elimination of infection

This is a very desirable aim but very difficult to achieve. In spite of the measures mentioned above, it is likely that infection will still persist, albeit at low levels.

The ultimate solution would be to cull the whole flock and restock with clean animals from EAE-tested flocks. This is unlikely to be a financially viable option, and there is always the nagging doubt of infection persisting on the farm, e.g. in housing, and infecting the clean stock.

A second option which may be possible on some farms is the 'two flock option', where the original flock is retained but clean replacements (from tested flocks) are maintained as a separate flock. As the years go by the original infected flock is progressively culled, and the new clean flock becomes 'the flock'. With a little flexibility the overlap can be limited to two years.

Great efforts must be made to keep the clean flock clean: lamb it first. Do not buy in lambs for adoption purposes – they may be infected.

Avoiding enzootic abortion of ewes

Enzootic abortion of ewes in a flock is a highly undesirable state of affairs. It is easily avoided, but very few sheep farmers seem to be interested in this – until they have found the infection in their flocks! Two methods are available: maintain a closed flock, or buy in replacements only from EAE-tested flocks.

The first option, a closed flock, is highly desirable from many disease points of view. 'A sheep's worst enemy is another sheep' may be something of an exaggeration but 'a sheep's worst enemy is another sheep of origin unknown' is most certainly not (Fig. 5.1). It is no coincidence that enzootic abortion of ewes is thankfully uncommon in self-replacing hill flocks.

If a closed flock is not possible, then only buy clean replacements. This may involve a change of traditional buying policy but the potential rewards are inestimable.

As mentioned above, don't buy in lambs for fostering. They may well be carrying enzootic abortion of ewes.

Figure 5.1   The new replacements – asset or liability?

Human infection
Note the contents of 'dangers to people' (p. 127), especially to pregnant women.

Toxoplasmosis

The cause
Together with enzootic abortion of ewes, toxoplasmosis is the other common cause of infectious abortion. But that is about all that toxoplasmosis has in common with enzootic abortion. In practically every other respect it is entirely different – save the end result of abortion and financial loss.

Toxoplasmosis is caused by a protozoan parasite, *Toxoplasma gondii*, related to the organism which causes coccidiosis in lambs. The organism infects the placenta, where it interferes with passage of nutrients and oxygen to the foetus. The infection also gains access to the foetus itself.

The disease
The source of infection for sheep, and indeed other animals including man, is the domestic cat (Fig. 5.2). When a non-immune cat, normally

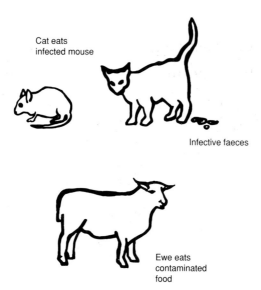

Cat eats
infected mouse

Infective faeces

Ewe eats
contaminated
food

Figure 5.2   Infection of ewes with *Toxoplasma* oocysts (eggs)

a young cat, ingests an infected mouse or infected uncooked meat it develops toxoplasmosis in its gut. About a week after infection the cat commences to pass literally millions of oocysts (eggs) in its faeces. This stage only lasts about a week, after which the cat becomes immune and passes practically no oocysts at all. It is these oocysts in cat faeces that infect sheep (and man).

This sequence of events contains two elements of importance in preventing disease in sheep. First it is mainly the young cat which poses a threat. The older immune cat is not a problem unless the cat itself is suffering ill-health. It would seem prudent to neuter the farm queens. Secondly, efforts must be taken to avoid contamination of sheep food and bedding with cat faeces. Straw and hay ricks are favoured nesting sites for pregnant queens, and an open grain store makes an excellent defaecation site – a sort of giant dirt tray!

There are two further important points:
1. Infection of the ewe, at any time including when she is not pregnant, leads to life-long immunity.
2. There is no transmission of infection from ewe to ewe, nor from tup to ewe. Toxoplasmosis is an infection of the farm, not the flock.

What are the effects of infection? Infection early in pregnancy can lead to foetal resorption and apparent barrenness. Infection in mid-pregnancy can lead to foetal death and abortion although a dead and often mummified foetus can be carried to term especially if a litter mate survives. Infection in late pregnancy can lead to the birth of weak lambs, or practically no ill-effects at all. When a non-immune flock meets toxoplasmosis for the first time the effects can be catastrophic with 30 per cent or more of ewes barren or aborting. In subsequent years, mostly non-immune replacement ewes will be affected.

Action in an outbreak

There are no precautions specific to toxoplasmosis, but until confirmation of the cause of the abortion you must assume that it is enzootic abortion of ewes. Therefore, follow the action already outlined for this disease (p. 116).

Future flock policy

Since ewes which have been infected are immune for life and pose no threat to other ewes they should be retained. This point emphasises the need for accurate diagnosis. If two ewes have aborted with enzootic abortion they should quite definitely be culled.

Avoiding toxoplasmosis

Neutering of female cats and avoiding contamination of feed and bedding for pregnant ewes have already been mentioned.

Recently, a new vaccine against toxoplasmosis, Toxovax, has been introduced. This one-dose live vaccine is effective in dramatically reducing losses from toxoplasmosis. The vaccine can be given to females aged five months and older, and should only be used in non-pregnant ewes. The vaccine should not be given in the three weeks preceding tupping.

Immunity after vaccination lasts at least two years but on farms with a toxoplasmosis problem natural infection should boost this to a life-long immunity. Thus on many farms vaccination of replacement females will be all that is required.

For safe and effective use this vaccine should be administered carefully according to the manufacturer's instructions. It is only available from your veterinary surgeon and you should contact him or her well in advance of anticipated use, i.e. months, for this vaccine is only made to order.

Monensin has been used to control toxoplasmosis. This drug does not kill the infectious agent but it does depress its activity, allowing more ewes to progress to term and thus to reduce the losses due to abortion. Monensin is given daily in feed. This itself poses a problem, for few sheep farmers wish to offer their ewes concentrates throughout pregnancy. Do not be tempted to overcome this problem by feeding a little concentrate with a high concentration of Monensin. If you do most ewes won't touch it but a few will consume it eagerly. The result, is a dead sheep – Monensin in excess is very poisonous.

Monensin is not licensed for use in sheep, nor is it likely to be so. It can only be obtained on veterinary prescription, and there is some considerable doubt as to whether such flock medication with unlicensed products will be permitted in the future. Deccox has also been found useful in the prevention of losses from toxoplasmosis.

The only cost-free preventative policy is to try and encourage infection in young stock before their first pregnancy. In this respect home-bred replacements probably have the advantage. In the first six months (breeding as ewe lambs) or 18 months (breeding as gimmers) of life they will meet *Toxoplasma* infection, and harmlessly acquire immunity. The replacement ewe lamb or gimmer bred on a hill farm, where toxoplasmosis is unusual, is unlikely to have met the infection.

All that remains is to try and encourage infection in non-pregnant stock: graze the flock near buildings occupied by cats, and allow the flock access inside the buildings contaminated by cats. This is fine in theory but seldom seems to work in practice.

Salmonellosis

The cause

Abortion can be caused by a variety of types of Salmonella bacteria. This group of bacteria can be divided into two classes: *Salmonella abortus ovis* and other salmonellae. Abortion due to *Salmonella abortus ovis* is now rare and mostly confined to south-west England.

Readers should be wary when consulting older texts on the subject of salmonella abortion. These accounts may well refer only to *Salmonella abortus ovis* disease, which differs in many important respects from other Salmonella infections.

Abortion due to *Salmonella abortus ovis*

As the name suggests, the sheep is the important species involved. *Salmonella abortus ovis* causes abortion in the last six weeks of pregnancy. The ewes show minimal effects and become immune for life, although they may well remain carriers of the infection. These ewes thus serve as a source of infection for clean stock when they abort, and for the remainder of their lives in the flock. Buying in a carrier ewe may well introduce the disease to a hitherto clean flock. This infection tends to become endemic within a flock, but use of a specially made vaccine can help to protect replacement ewes.

Disease caused by other salmonellae

A variety of Salmonella bacteria can cause abortion. The list includes *Salmonella dublin* and *Salmonella typhimurium* which also causes disease in man, and many other exotic types including *Salmonella montevideo*.

Infection is by mouth and up to 20 per cent of the flock may abort. The source of infection varies and includes other stock, e.g. calves, contaminated feed or water, sewage and wild birds. Symptoms in affected ewes vary with the type of Salmonella involved. *Salmonella montevideo* produces no symptoms save abortion, *Salmonella dublin* produces scour and fever and *Salmonella typhimurium* causes severe illness, including death. In some *Salmonella typhimurium* outbreaks dead ewes, not abortion, is the first sign. Some salmonellae can cause serious disease in man. Shepherds and their families are at special risk.

Action in an outbreak

Aborting ewes are a threat to other sheep and to man. All the precautions outlined for enzootic abortion should be followed (p.116). Every abortion should be regarded as being caused by enzootic abortion and salmonellosis, i.e. a serious threat to sheep and people, until it is proved otherwise.

If Salmonella is demonstrated to be the cause of abortion, your veterinary surgeon may advise antibiotic therapy for the whole flock.

Future flock policy

In many cases Salmonella abortion is a one-off event and serious problems do not occur in the future. Culling of aborting ewes is seldom advised.

A number of Salmonella vaccines are available, but their potential usefulness depends on the type involved. Your veterinary surgeon will be able to advise (p.143).

Avoiding salmonellosis

Most sources of salmonellae will be unknown until an outbreak, and little can be done to eliminate them. Infected calves are a known source and if Salmonella occurs in calves on a sheep farm every effort must be made to avoid cross-infection. Ideally, staff tending sheep should have no contact with infected calves. Sheep should have no access to contaminated buildings.

Wild birds can be another potential source of infection. The most likely contact point is outside feeding troughs. It makes sense to clean these regularly, to upturn them after the sheep have fed and to move the troughs daily.

Campylobacteriosis

The cause

Campylobacteriosis (formerly known as vibriosis, a much more pronounceable name!) is caused by two species of the bacteria campylobacter, *Campylobacter foetus* and *Campylobacter jejuni*. This latter organism can cause disease in man.

In contrast to cattle, campylobacteriosis in sheep is not a venereal disease. It cannot be passed from the tup to the ewe.

The disease

Infection is by mouth and infection after three months of pregnancy causes abortion within 1–4 weeks. Infection before this time or when the ewe is not pregnant causes no serious disease, and confers life-long immunity. The source of infection may be a bought-in carrier ewe, but in some outbreaks birds or rodents may be responsible.

The aborting ewe is a rich source of infection. The foetus, placenta and discharges are heavily contaminated. Other ewes which pick up the infection in late pregnancy may produce stillborn lambs but some will produce weak live lambs. The aborted ewe is solidly immune and will not abort again.

Action in an outbreak

Apply the same measures as for enzootic abortion (p. 116). Don't forget that the infection might cause disease in people.

Your veterinary surgeon may advise antibiotic therapy for remaining pregnant ewes.

Future flock policy

Fortunately campylobacteriosis tends to be a one-season affair, serious problems not occurring in future years. Aborted ewes should be retained.

If it is thought necessary a special vaccine may be prepared for use in replacement ewes.

Avoiding campylobacteriosis
The threat of the infected bought-in ewe has already been mentioned. Breeding your own replacements avoids this risk.

Listeriosis

The cause
Two types of listeria bacteria cause disease in sheep, *Listeria monocytogenes* and *Listeria ivanovii*. This is a ubiquitous organism found in the soil and elsewhere, though normally in small numbers. The incidence of listeriosis has increased considerably in recent years paralleling the increased feeding of silage to sheep. There is no doubt that poor silage is the major source of infection for sheep.

Well-fermented silage with a pH value below 5 does not permit the multiplication of listeria bacteria, but poor silage which is less acid does. This poor silage is the major source of infection.

Listeria infection does not only cause abortion in sheep. It also causes encephalitis in adult sheep – circling disease – and a serious septicaemia in lambs (p.63).

The disease
Abortion occurs a few weeks after infection with the production of dead or weak lambs. In the absence of ewes affected by the encephalitic form of listeriosis, abortion is the first sign seen. Although the products of abortion are contaminated it is not generally thought that sheep-to-sheep transmission is of great importance. Heavily contaminated food would appear to be much more important.

The incidence of listeria abortion in a flock is normally low – around the 1 per cent mark, but occasionally 10 per cent of ewes may abort.

Action in an outbreak
Proceed as for enzootic abortion (p. 116). If *Listeria* is suspected the silage should be examined. However, do not be surprised if a satisfactory analysis is found. The important silage was that consumed three weeks ago.

Future flock policy
The ubiquitous nature of the organism means that culling of aborted ewes is pointless.

Avoiding listeriosis

Attention should be paid to silage quality and handling. Ensure that pH value is below 5 (acid), and ash content below 70 parts per million (minimal soil contamination).

With clamp silage the surface layers are the most likely to be contaminated and if possible they should not be fed to sheep. Cattle are less susceptible to listeriosis than sheep and it is often suggested that spoilt or dubious silage should be fed to them. This must still entail a small risk.

Do not allow uneaten silage to lie in feeding passages. Exposure to the air will allow the pH value to rise and facilitate multiplication of *Listeria* bacteria.

Big bale silage from damaged bags should not be fed to sheep.

Border disease

The cause

Border disease is caused by a virus closely related to the virus causing bovine virus diarrhoea in cattle. In fact, sheep can acquire infection from cattle. As the name suggests the disease was first identified in the Welsh border counties, but now it is found throughout the United Kingdom and elsewhere.

The disease

In non-pregnant stock infection causes very mild disease leading to life-long immunity, but in pregnant ewes infection can lead to various problems including foetal resorption and apparent barrenness, abortion or the birth of weak lambs. Some live lambs show the typical 'hairy shaker' syndrome – excessive shaking and a hairy coat. Whilst the hairy shaker lambs are often seen in outbreaks, this is not always so.

The source of infection is a bought-in carrier ewe or tup. It is vital to understand the origin of such an animal.

If infection occurs in late pregnancy, the foetus reacts from an immunological point of view like an adult – it mounts an immune response producing antibodies. These antibodies eliminate the virus. The foetus may be stillborn, born weak, show the hairy shaker syndrome or be apparently normal, but it will not be infected.

However, if infection occurs in early pregnancy, the foetus mounts no immune response; it accepts the virus as 'self'. Such a foetus may not survive, but if it does it will be born with no antibodies but, of great importance, persistently infected with virus – a symptomless carrier.

Action in an outbreak

There may be no signs at abortion that the cause is obviously border disease. Therefore it must be assumed that the abortion is caused by enzootic abortion of ewes and salmonellosis, and all the precautions noted for enzootic abortion observed (p. 116).

Future flock policy

There are two possible approaches. The first is to eliminate the disease, and the second to live with it.

To eliminate the disease the carrier animals must be identified. Blood sampling of all stock will be required. This approach may be possible in a small flock or in a larger flock where a group of suspect animals can be identified.

If the elimination approach is not possible every effort must be made to encourage infection when the ewes are not pregnant, up to two months before tupping. Ewes and lambs must be run together, ideally under some close confinement. Border disease is not a particularly contagious disease, and extensive grazing conditions would be unlikely to serve this purpose. Housing at night might be an option.

Avoiding border disease

The bought-in carrier animal is the problem. The closed flock is at much reduced risk. In open flocks every effort should be made to ensure that replacements come from border disease-free flocks. The closed flock is not totally safe. A purchased carrier tup is a potential risk. Newly acquired tups should be isolated on arrival at the farm, and blood tested to ensure freedom from border disease.

Other infectious causes of abortion

Any infection in pregnancy causing fever and malaise, even if for only a few days, can cause abortion. One of the best known examples is tick-borne fever.

Ewes bred in tick country encounter the disease as lambs and gain life-long immunity but if tick-borne fever-naive pregnant ewes from a tick-free area are moved to a tick area they will become infected, and some will abort. Prevention is common sense – don't do it.

The organism *Coxiella burnetii* is carried without symptoms by many sheep but occasionally it causes abortion. This organism causes Q fever in man and hence the importance once again of following hygienic procedures at lambing.

Brucella abortus, more commonly associated with cattle, can cause abortion in sheep. With the eradication of this disease, problems in sheep have hopefully disappeared.

Mouldy hay can occasionally cause abortion. The mould produces toxins which cause the abortion.

Avoiding problems in people

A glance at Table 5.2 shows that many infectious causes of abortion can cause disease in people.

As always, the very young and very old are at increased risk. But just as the pregnant sheep suffers serious consequences when meeting these infections (but little disease when not pregnant), so also can pregnant women. The pregnant uterus would appear to provide a safe haven for many of these infections.

The moral is very simple: pregnant women have no place in lambing situations and should keep well away.

To avoid risk to lambing personnel remember that until proved otherwise, all abortions are caused by enzootic abortion and salmonellosis. Aborted ewes are thus a potential danger to people and to other sheep. Observe these common-sense precautions:

1. Use disposable gloves especially when handling the products of abortion.
2. Wash hands thoroughly before eating, drinking or smoking.
3. If there is a pregnant woman in the household, discard outer clothing before entering the house.
4. Isolate all aborted sheep.
5. Bury or burn all placentae and foetuses.

Avoiding infectious abortion

Infectious abortion is a dismal subject, miserable to write about and equally miserable to read about. From a veterinary point of view there is seldom a magic answer to the problem.

Keeping the flock clean must be the aim. Some problems we cannot prevent, e.g. infected wild birds, but most potential problems can be prevented.

With many infections (e.g. enzootic abortion) the bought-in infected ewe is the culprit. Ideally breed your own replacements. If this is not possible buy from known enzootic abortion-tested flocks. Buying at big sales may be an enjoyable social occasion but it is a lottery, a bit like playing poker with a blindfold (Fig. 5.1).

Most flocks purchase tups and here the risk is border disease. The tups must be blood-tested.

Chapter 6

The prevention of problems

Some problems cannot be foreseen and must be treated as and when they arise. Most problems, however, can be prevented and in this chapter we have summarised preventative measures which are applicable to most flocks. We also outline how problems at lambing can be accurately assessed and a prevention programme tailored to the individual farm.

Prevention of problems in ewes

For further details of the specific problems mentioned below (in parentheses) see Chapters 3 and 4.

1. DO condition-score ewes regularly through pregnancy and give extra feeding to lean ewes. In lowland flocks, aim for a condition score of 3–3½ at tupping, 2½–3 at mid-pregnancy and 3 at lambing. For the hill flock, the scores might be 2½–3, 2½ and 2½–3 (pregnancy toxaemia and practically all lamb problems).
2. DO NOT impose any sudden change in nutrition (pregnancy toxaemia).
3. DO NOT feed poor quality roughage (pregnancy toxaemia, prolapses).
4. DO feed the daily concentrate ration in two feeds (pregnancy toxaemia, prolapses).
5. DO feed a high magnesium concentrate if hypomagnesaemia is a problem in your flock.
6. DO vaccinate your ewes against the clostridial diseases (metritis in ewes, lamb dysentery, pulpy kidney and tetanus in lambs).
7. DO NOT stress ewes in late pregnancy or after lambing (hypocalcaemia).

8. DO observe all the warnings given in Chapter 2 when assisting a lambing (metritis, physical injury).
9. DO give long-acting antibiotic to ewes after an assisted lambing (metritis).
10. DO NOT let hungry lambs butt an empty udder (mastitis).

Prevention of problems in lambs

For further details of the specific problems mentioned below (in brackets) see Chapters 1, 3 and 4.

1. DO condition-score ewes regularly through pregnancy and give extra feeding to lean ewes. In lowland flocks aim for a condition score of 3–3½ at tupping, 2½–3 at mid-pregnancy and 3 at lambing. For the hill flock, the appropriate scores might be 2½–3, 2½ and 2½–3 (stillbirth, prematurity, low birth weight, poor body energy reserves, low colostrum production in the ewe leading to starvation and little resistance to infectious disease in the lamb).
2. DO vaccinate ewes against clostridial disease and ensure that lambs suck plenty of colostrum (lamb dysentery, pulpy kidney, tetanus).
3. DO provide shelter if lambing outside (hypothermia).
4. DO ensure adequate labour during lambing. Tired bad-tempered shepherds make mistakes.
5. DO dry lambs after birth if the ewe fails to do so, especially small twins and triplets (hypothermia).
6. DO dress navels as soon as possible after birth (joint ill, liver abscess, navel ill, spinal abscess).
7. DO clean and disinfect lambing pens after every ewe (ALL infectious diseases).
8. DO ensure that lambs get plenty of colostrum within a few hours of birth – give by stomach tube if necessary (hypothermia, ALL infectious diseases). Consider penning lambing ewes before they lamb rather than 30 minutes after lambing, when the lambs are starting to suck.
9. DO detect and treat entropion as early as possible.
10. DO detect hungry lambs – temperature check, and supplement by stomach tube (hypothermia).
11. DO NOT feed weak lambs by bottle (inhalation pneumonia).
12. DO NOT castrate lambs with rubber rings before 12 hours of age. This reduces colostrum intake and makes watery mouth more likely.

13. DO detect hypothermia in the early stages (temperature check) and treat quickly.
14. DO NOT turn out lambs which are hungry (temperature check). They will become hypothermic.
15. DO NOT keep lambs on wet bedding or sodden pasture (scad).

Preventing lamb problems

All sheep farmers and shepherds recognise the triplet or small twin lamb as a poor risk, less likely to survive than the fit single.

However, our approach to these problem lambs seems to be one of passive pessimism – wait and see what happens. When it is clear that something is amiss, the resuscitation brigade is called for, often to no avail.

We need an active approach to these lambs aimed at preventing problems. First spot the lambs. The list includes:
1. All triplets.
2. Twins out of thin ewes.
3. Twins out of ewe lambs.
4. Lambs which appear weak at birth.
5. Premature lambs, e.g. abortion.

What do these lambs need in the first few hours of life? The first requirement is drying. If the ewe has not completed this task – *you* must. A towel is infinitely more effective than a handful of straw!

Secondly, food. Ideally use stored ewe colostrum but if not available use cow colostrum or a substitute. There is a slight risk of anaemia with cow colostrum, but cow colostrum has saved many more lambs than it has killed. Give about 150 ml feed by stomach tube. The aim of this first feed is to give the lamb the strength to get to its feet and to suck. Feeding by stomach tube encourages sucking, not the reverse as is sometimes thought. By three or four hours of age the lambs should be lying with the ewe, contented and full, not struggling around the pen still trying to find the teat.

Assuming this happy state has been established, the lamb's future depends on a continual supply of milk. Supplementation may be required. When supplementing healthy lambs the bottle can prove useful, giving some idea of the deficit. Remember to feed all the lambs, not just the one which is obviously hungry.

Assessment of losses

There is a tradition in sheep farming that lambs are never counted until after lambing. This tradition is based on the idea that 'if you never had it, you can't have lost it'. In the days before the advent of clostridial vaccines and antibiotics many lamb losses could not be prevented and this tradition is very understandable. But it has no place in modern sheep farming. It is essential that the flockmaster knows how many ewes and lambs have died, why they have died and, equally important, what factors predisposed to the deaths. You must keep records.

Before lambing, discuss plans for lamb and ewe death recording with your veterinary surgeon, who will be largely responsible for the interpretation of your records. Time and effort will be needlessly wasted if you do not involve the vet at the planning stage. You, however, will have to do the bulk of the work.

The first question to be answered is, 'Do I have a lamb or a ewe mortality problem?' To answer this question, simple records need to be kept to enable a flock performance survey to be completed (Table 6.1). There must be no exceptions to this recording – all deaths must be recorded. All lambing staff must understand the aims of the exercise: the shepherd must not be deterred from recording losses by fear of recriminations from the boss!

The results of the recording exercise must be examined in the cool light of day after lambing. The figures below give our interpretation of lamb and ewe losses.

Lamb losses (percentage of all lambs born)

0 per cent:	Impossible.
1 per cent:	You are deceiving yourself.
5 per cent:	The best mean figure that the authors can achieve.
10 per cent:	A good figure for the average commercial flock. Room for some improvement.
15 per cent:	Average. Aim for a reduction to 10 per cent.
20 per cent:	Too high. Improvements in the correct area will yield gratifying results.
25 per cent	(or more): Much too high. Either improve or get out of sheep farming!

Ewe losses (percentage of ewes lambing)

1 per cent:	Good.
2 per cent:	Average. Pinpoint the major problems and attempt to prevent them.
3 per cent	(or more): You have a problem. Detailed investigation is required.

Table 6.1 Flock lambing performance record

Ewes lambing			.....
Ewes barren			.....
Lambs born (dead and alive)		singles	.....
		twins	.....
		triplets	.....
			———
		total	———
Lambs dead within 7 days of birth			
(including stillbirths)		singles	.....
		twins	.....
		triplets	.....
			———
		total	———
Ewes dying			.....

By calculation

$$\text{Lambing (\%)} = \frac{\text{lambs born}}{\text{ewes lambing}} \times 100$$

Lamb mortality in first 7 days (%)

$$\text{Singles} = \frac{\text{dead singles}}{\text{singles born}} \times 100$$

$$\text{Twins} = \frac{\text{dead twins}}{\text{twins born}} \times 100$$

$$\text{Triplets} = \frac{\text{dead triplets}}{\text{triplets born}} \times 100$$

$$\text{Total} = \frac{\text{dead lambs}}{\text{lambs born}} \times 100$$

Ewe mortality (%)

$$\frac{\text{ewes dying}}{\text{ewes lambing}} \times 100$$

If examination of the figures suggests that losses can be reduced, further recording may be necessary to pinpoint the areas of management requiring improvement (Tables 6.2 and 6.3). Blank lamb and ewe records can be duplicated onto loose sheets or, more conveniently, can be made up into a rubber stamp and stamped into the pages of a pocket notebook. The information gained from these records is itself most useful, but its value will be enhanced if post-mortem examinations are also performed. Discuss this with your veterinary surgeon.

At the end of lambing, pass your records to your vet for interpretation. The information gained from the records together with the results of post-mortem examinations will enable the vet to pinpoint the causes of death and important predisposing factors in most cases. A few examples may help to show how this can be done.

Example 1. Lamb
Stillborn, single, 6.0 kg, birth assisted, fresh carcass.
Ewe: 2 years old, condition score 3, no disease.
Post-mortem: lungs not inflated, no other findings.
Probable cause of death: parturient stillbirth.
Predisposing factors: big lamb, maiden ewe, assistance too late?

Example 2. Lamb
Died at 48 hours of age, twin, 3.5 kg, hypothermic, other twin also hypothermic.
Ewe: 4 years old, condition score 1½, no disease.
Post-mortem: empty stomach and intestines, fat reserves exhausted, no other findings.
Probable cause of death: hypothermia due to starvation.
Predisposing factors: thin ewe, little milk, poor nutrition during pregnancy.

Example 3. Lamb
Destroyed at 6 days of age, twin, 4.5 kg, off back legs for last 2 days, other twin healthy.
Ewe: 3 years old, condition score 3, no disease.
Post-mortem: abscess pressing on spinal cord.
Disease: spinal abscess.
Predisposing factors: dirty pen? navel dressed?

Table 6.2   Lamb death record

Date	Time ....

Weather in last 12 hours (especially if outside)

....

Type   single/twin/triplet

Age at death   stillborn/0–5/5–12/12–24/24–48 hours

3–5/5–7 days

Assisted birth   Yes/No

Weight   .... kg

If stillborn   fresh/decomposed

Fate of other lambs (if a twin or triplet) ....

Symptoms   ....

Ewe age   .... years

Ewe condition score   ....

Ewe disease   ....

Evidence of abortion   Yes/No

Table 6.3   Ewe death record

Time of death   before lambing/at lambing/after lambing ..... days

Age   .... years

Condition score   ....

Assisted at lambing   Yes/No

Lambs born   ....

Symptoms   ....

Example 4. Ewe
Died 3 days after assisted lambing, 3 years old, condition score 3, foul-smelling discharge from vulva.
Cause of death: metritis.
Predisposing factors: poor hygiene at lambing, long-acting antibiotic given?

Example 5. Ewe
Died 5 days after lambing triplets, 5 years old, condition score 2, high temperature, one side of udder swollen and painful.
Cause of death: mastitis.

Predisposing factors: poor nutrition during pregnancy, insufficient milk for three lambs.

As an illustration of how recording at lamb can help to improve flock performance we have presented in Table 6.4 a summary of the lamb death situation in one commercial flock. The mortality rate in this flock was too high (18 per cent). The major cause of loss was hypothermia due to starvation. Poor ewe nutrition, reflected by low condition scores, was the important predisposing factor. Not surprisingly, the mortality rate in triplets was very high (28 per cent). Other significant causes of death were infections (probably related to a low colostrum intake), parturient stillbirths and foetal stillbirths, probably related to poor nutrition.

Table 6.4 Summary of the lamb death information gained from one commercial flock.

Flock performance			
	Ewes lambing	416	
	Total lambs born	818	
	Lambing percentage	197%	
Lamb mortality			
	Singles	8	(10% of all singles born)
	Twins	79	(15% of all twins born)
	Triplets	56	(28% of all triplets born)
	Not known	3	
	Total	146	(18% of all lambs born)
Causes of lamb death			
	Stillbirth – foetal	13 lambs	(9%)
	Stillbirth – parturient	13 lambs	(9%)
	Hypothermia – exposure	19 lambs	(13%)
	Hypothermia – starvation	59 lambs	(40%)
	Infectious disease	28 lambs	(19%)
	Other problems	5 lambs	(4%)
	No diagnosis	9 lambs	(6%)
	All causes	146 lambs	(100%)
Predisposing factor			
	Condition scores of the ewes which lost lambs		
	Condition score 1	78 ewes	
	Condition score 2	15 ewes	
	Conditions score 3	4 ewes	
	Not recorded	24 ewes	

The following changes in management were recommended:
1. Improve ewe nutrition. Introduce regular condition scoring and draw out lean ewes for extra feeding. Consider ultrasonic scanning for the identification of twin-and triplet-bearing ewes.

2. Upgrade the standard of hygiene at lambing – lambing pens, navel dressing, assisted lambings.
3. Send the shepherd and other farm staff on Agricultural Training Board courses on hypothermia and lambing.
4. Improve triplet management, especially nutrition.

Summary

Most lamb and ewe losses are preventable and this should be your aim. Remember that ewe and lamb losses represent only one part of the true loss. For each lamb that dies another 'just makes it' and suffers a severe check to its growth and development. For each ewe that dies another may be successfully treated but performance in lactation is likely to be poor and the lambs held back. Conditions such as mastitis may result in premature culling.

Lamb performance recording, the correct interpretation of the findings and the implementation of appropriate improvements are likely to have dual benefit: losses will be reduced and the productivity of the whole flock will be increased.

Ewe body condition scoring

Throughout this chapter and elsewhere we have stressed the importance of ewe nutrition and body condition. In this section we outline how ewe body condition can be objectively assessed using the body condition scoring technique. An objective system for assessing body condition is needed for two reasons:

1. Definitions of ewe condition such as 'poor, lean, fat, moderate, fit and good' vary from individual to individual.
2. The individual's definition of these descriptions tends to vary from one year to another depending on the general nutritional state of the flock. The definition of 'fit' at lambing in a good year is unlikely to be the same as 'fit' in a poor year.

Ewes are scored on a scale ranging from 0 to 5 using half scores when needed to improve accuracy. The score is related to the degree of fatness in the lumbar region of the back, behind the rib-cage (Fig. 6.1). The score is assessed in four stages (Fig. 6.2):

1. Assess the degree of prominence of the spinous processes of the lumbar vertebrae.

Score		Description	
1		Spine sharp, back muscle shallow, no fat	Lean
2		Spine sharp, back muscle full, no fat	
3		Spine can be felt, back muscle full, some fat cover	Good condition
4		Spine barely felt muscle very full, thick fat cover	
5		Spine impossible to feel, very thick fat cover, fat deposits over tail and rump	Fat

Figure 6.1 Ewe body condition scoring (see text for details). (From Speedy, 1980)

2. Assess the prominence and degree of fat cover over the ends of the transverse processes.
3. Assess the degree of muscle and fat cover beneath the transverse processes by judging the ease with which the fingers may be passed under these bones.
4. Assess the fullness of the eye muscle and fat in the angle between the spinous and transverse processes.

Once you have completed your examination, score the ewe according to the scale overleaf using half scores when needed:

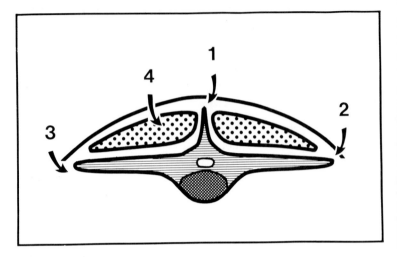

Figure 6.2   Ewe body condition scoring. A cross-section through the
lumbar spine. (After Russel, 1984)

Score 0

Extremely emaciated and on the point of death. It is not possible to
detect any muscular or fatty tissue between the skin and the bone.

Score 1

The spinous processes are prominent and sharp; the transverse
processes are also sharp, the fingers pass easily under the ends, and it
is possible to feel between each process; the loin muscles are shallow
with no fat cover.

Score 2

The spinous processes are still prominent but smooth, and individual
processes can be felt only as fine corrugations; the transverse processes
are smooth and rounded and it is possible to pass the fingers under the
ends with a little pressure; the loin muscles are of moderate depth, but
have little fat cover.

Score 3

The spinous processes have only a small elevation, are smooth and
rounded, and individual bones can be felt only with pressure; the
transverse processes are smooth and well covered, and firm pressure is

required to feel over the ends; the loin muscles are full, and have a moderate degree of fat cover.

Score 4

The spinous processes can just be detected with pressure as a hard line, the ends of the transverse processes cannot be felt; the loin muscles are full, and have a thick covering of fat.

Score 5

The spinous processes cannot be detected even with firm pressure; there is a depression between the layers of fat in the position where the spinous processes would normally be felt; the transverse processes cannot be detected; the loin muscles are very full with very thick fat cover. There may be large deposits of fat over the rump and tail.

With a little practice you will be able to score ewes quite quickly. Regular use of this technique will enable you to rationalise the nutrition of your ewes and will ensure that extra food is given to the ewes which need it.

Vaccination

The days are long gone when sheep farmers could afford to vaccinate their ewes against everything, 'just in case'. Expenditure on vaccination must be justified in the light of previous experience.

The only exception to this statement is vaccination against the clostridial diseases such as lamb dysentery, pulpy kidney and tetanus. These are horrific diseases and can be prevented by using the cheap effective vaccines available. A relaxation in vaccination policy by some sheep farmers in recent years has resulted in an increased incidence of these diseases. Failure to use these vaccines is tantamount to negligence. When using any vaccine it is essential to study the instructions supplied. Recommendations do change with experience and vaccine development.

Clostridial vaccines

Clostridial vaccines are presented as multivalent vaccines covering a variety of combinations of diseases. These diseases include lamb dysentery, struck, pulpy kidney, braxy, blackleg, tetanus, black disease, bacillary haemoglobinuria and post-parturient gangrene. The widest vaccines include protection against all these diseases but other products are restricted, e.g. braxy, blackleg, pulpy kidney and tetanus. All these products are dead vaccines and thus two doses are required to confer protection. One dose confers little or no protection.

Two forms of protection are acquired from clostridial vaccination. First the ewe or ram is protected from the diseases affecting principally adult animals, e.g. black disease. Secondly, through the colostrum the newborn lamb is protected from killers such as lamb dysentery, pulpy kidney and tetanus. This immunity in the lamb lasts about three months. Thereafter the lamb itself must be vaccinated if protection is to be continued.

Ewes

Ewes require two doses of vaccine separated by 4 – 6 weeks. To ensure protection for newborn lambs, ewes should receive a booster dose at least one month before lambing is due to commence.

Annual revaccination before lambing is normally practised, but experience of autumn or winter disease in ewes may indicate twice yearly boosting.

Lambs

Lambs from unvaccinated ewes can be vaccinated from two weeks of age, but lambs from vaccinated ewes should be protected for the first 2–3 months of life. Vaccination of these lambs can commence at two months of age.

Erysipelothrix vaccine

Erysipelothrix infection causes joint ill in lambs and post-dipping lameness in older sheep.

The vaccine is a dead preparation and thus two doses are required, 2–6 weeks apart. If the major infection problem occurs in young lambs the ewes will require vaccination, the second or booster dose being given a month before lambing is due to commence. Immunity contained in the colostrum should protect the lambs.

If the major infectious problem occurs later in life (e.g. at dipping), the lambs and ewes will require vaccination. The second or booster dose should be given at least two weeks before the risk period.

Escherichia coli *vaccine*

This dead vaccine is specifically formulated for the prevention of enteritis in lambs caused by *E. coli*. Ewes are given two doses at least two weeks apart, the second dose being given one month before lambing is due. In subsequent years only a single booster is needed one month before lambing.

Foot rot vaccine

Foot rot is an uncommon problem in young lambs, but it can be a severe problem in ewes. Disease in ewes has two significant consequences: pain and distress for the ewe, and secondly poor nutrition which will affect foetal development and colostrum production. From all points of view this is a disease to control, and if possible eradicate.

Vaccination alone is unlikely to be successful in controlling foot rot, but in conjunction with other measures such as foot bathing and the culling of chronically infected ewes control is possible. In closed flocks the control should be followed by eradication.

Foot rot vaccines are dead preparations and the dosage regime depends on the particular vaccine used. After initial vaccination boosters are recommended to precede risk times, generally when conditions are wet underfoot.

Take note of the instructions accompanying the vaccine. One of the products contains an oil and causes severe reactions if self-injection occurs. Immediate medical attention at a hospital accident/emergency department is required.

Louping ill vaccine

Use of louping ill vaccine is restricted to areas where the disease is endemic. The vaccine is a dead preparation suspended in an oil emulsion. Only one dose is required.

Sheep must be vaccinated at least one month before exposure to infection. Lambs from vaccinated ewes will be protected for the spring risk period by colostral immunity, but they will require vaccination in late summer to protect them from autumn challenge.

Tick-naive sheep must be vaccinated one month before introduction to tick pasture.

Revaccination at two yearly intervals is recommended, but this may be unnecessary in endemic areas where stock are repeatedly subject to natural challenge.

As with other oil-based vaccines, accidental self-injection can have serious consequences and urgent attention at the nearest hospital accident/emergency department is required.

Orf vaccine

Orf vaccine is live orf virus and thus this product should never be used on farms free of this disease.

Orf is a skin disease and the vaccine is used to create the disease at a skin site which will not unduly inconvenience the sheep. The infection is followed by a degree of immunity lasting a few months. The vaccine is applied by scratching the skin with a special applicator carrying a drop of vaccine. Localised active orf is the result.

The resulting scabs which eventually fall to the ground contain live orf virus. If maintained in a dry environment (e.g. a sheep house), these scabs can remain infective to other sheep for many months. However, the virus has little resistance to the elements, especially water, and will soon perish if deposited on pasture.

This persistence of virus in dry scabs has serious implications for the use of orf vaccine. If orf is a problem in young lambs and nursing ewes, the ewes if maintained outside may be vaccinated not less than eight weeks before lambing.

If the ewes are housed in the future, lambing quarters vaccination may be contraindicated due to the persistence of infection in dry scabs. No immunity passes from the ewe to the lamb in colostrum, and thus lambs should be vaccinated soon after birth, after the birth coat has dried.

Vaccination of lambs only or ewes only carries the risk of exposing unprotected stock to infection.

If orf is only a problem in older weaned lambs vaccination of ewes may be unnecessary. The lambs should be vaccinated 3–4 weeks before the expected risk period.

The site of vaccination is chosen to minimise the risk of spread to other stock. In pregnant ewes either the base of the tail or behind the shoulder is chosen. The latter site carries a lower risk of secondary bacterial infection. In young lambs the axilla (armpit) is the best site, whereas in older lambs the inside of the thigh is chosen.

Follow the instructions supplied with the vaccine carefully. If the vaccine does not take, i.e. no active infection is caused, it will have no beneficial effect whatsoever. The pattern of orf infection and the apparent efficacy of orf vaccine seems to vary from farm to farm, even from year to year. Whether vaccination is practised at all, and the precise regime to follow, will depend very much on local conditions and previous experience.

Pasteurella vaccine

Dead pasteurella vaccines are available alone or combined with clostridial vaccines. *Pasteurella* bacteria cause both pneumonia and septicaemia in sheep. The vaccines are aimed principally against pneumonia. Being dead vaccines two doses are required, about one month apart. The second or booster dose should precede any risk period by at least two weeks.

On farms with a low incidence of *Pasteurella* pneumonia annual boosting at least one month before lambing should suffice. However if experience dictates, boosting at six-monthly intervals may be required.

Some immunity is transferred to the lamb in colostrum but this wanes in 3 – 4 weeks. Consequently, this is the age when vaccination of lambs can start, but on some farms serious disease in lambs can occur about this time. In these circumstances it is sometimes advised that initial vaccination of lambs should start in the first few days of life.

Salmonella vaccine

Salmonella vaccination is not routinely practised in sheep, but may be indicated where abortion due to salmonellosis has occurred, or it is known that Salmonella infection is present in other animals, e.g. cattle on the same farm.

The potential usefulness of these vaccines depends on the type of Salmonella involved. Take professional advice before considering using them. Salmonella vaccines are dead vaccines and two doses at least two weeks apart are required. This factor severely limits the usefulness of these products when faced by an outbreak. First, considerable spread of infection will have occurred before a diagnosis is made, and secondly no useful immunity will be gained from vaccination until 3 – 4 weeks after the first dose.

Vitamin supplementation

Vitamin supplementation is only of benefit when a deficiency exists. Supplementation to ewes or lambs suffering no deficiency is a waste of money. The B vitamins, and vitamins K and C, are synthesised in the rumen and no supplementation is normally required. The only exception to this is vitamin B_1 (thiamine), a deficiency of which causes cerebrocortical necrosis (p. 30). In contrast, the fat-soluble vitamins,

A, D and E, must be supplied in the diet. Deficiency in summer, when fresh herbage is available, is rare but when stock are dependent on preserved forage in the winter problems can arise.

Ewes

Good hay and well-preserved silage should be adequate in vitamins A, D and E, but poorly conserved forage is often deficient in these fat-soluble vitamins. Not surprisingly, vitamin deficiency often manifests itself in the spring following a bad summer. Bought-in concentrates and home mixes with a vitamin/mineral supplement should be adequate in vitamins A, D and E, but often these feeds are only fed in late pregnancy.

When past experience indicates a potential deficiency of these vitamins, or when it is known that preserved forage quality is poor, a combined vitamins A, D and E injection should be given to ewes at mid-pregnancy.

Lambs

Lambs from deficient ewes will have low body reserves of vitamins A, D and E, and will receive a less than adequate supply in colostrum and milk. The most likely problem is stiff lamb disease (p.87), which may be caused by vitamin E deficiency, selenium deficiency or both these circumstances. A deficiency of vitamin A will lead to high susceptibility to infectious disease, and a deficiency of vitamin D will retard skeletal development.

When experience suggests a potential problem with a vitamin deficiency lambs should be injected with a combined A, D and E preparation soon after birth. This should protect them until grazing of fresh herbage ensures an adequate supply.

In cases of stiff lamb disease, it is vital that a thorough investigation is conducted to determine the exact cause: vitamin E deficiency, selenium deficiency, or both these. Vitamin supplementation will have minimal benefit if selenium deficiency is the main problem, and vice versa.

Mineral supplementation

Magnesium

Magnesium deficiency is a problem of ewes only, leading to hypomagnesaemia, i.e. grass staggers (p. 103). It commonly occurs at turn-out onto flush pasture.

Extra magnesium must be supplied in the diet. A high magnesium concentrate will serve the purpose, but this may be economically undesirable when plentiful grass is available. Calcined magnesite may be spread on the pasture to raise the magnesium content of the sward. The practicality and efficacy of this method will vary considerably from farm to farm. Take advice from your agricultural adviser.

Probably the surest method to prevent magnesium deficiency without feeding a high magnesium concentrate is to use the magnesium 'bullet' (Fig. 6.3). Magnesium bullets are moulded metal cylinders composed of a magnesium alloy. The sheep bullet measures 19 x 46 mm, and is administered by means of a special balling gun. After administration, the bullet lies in the reticulo-rumen and releases magnesium over a period of three weeks or sometimes longer.

Free-access minerals such as powder or blocks are sometimes used in an attempt to prevent magnesium deficiency. Unfortunately, ewes vary enormously in their appetite for these products, and results are often very variable.

Copper

Copper deficiency can be a problem in both ewes and lambs (pp. 42 and 88). Sometimes supplementation of ewes seems to resolve any problem in lambs, but in some circumstances supplementation of both ewes and lambs is required.

Figure 6.3   A magnesium 'bullet'

The toxicity of copper to sheep has already been outlined (p. 43), and great care must be taken with copper products. DO NOT use these products unless you have hard evidence of deficiency, i.e. blood samples taken by your veterinary surgeon. Never use more than one form of copper supplementation. Take extra care if ewes are to be housed for any period, since the availability of copper in conserved forages considerably exceeds that from fresh grass. Note also that some breeds (e.g. Suffolk) are particularly susceptible to copper poisoning.

There are three types of copper supplement available: injections, copper oxide needles and the glass bolus which also contains cobalt and selenium (see below).

Copper injections

Ewes are generally injected around mid-pregnancy. Precise injection method varies with the product – read the instructions.

Lambs will acquire some copper in the ewe's milk. Thus injection of lambs is normally contraindicated until at least 6 weeks of age.

Copper oxide needles

Copper oxide needles are a novel form of copper supplementation. The animal is dosed with a capsule (or capsules) containing copper oxide needles (Fig. 6.4). These needles lodge in the wall of the abomasum whence copper is slowly released. One treatment normally provides adequate copper for a whole year. Ewes should ideally be dosed at mid-pregnancy but in many circumstances dosing as early as tupping will be adequate to prevent swayback. Make sure the capsule is swallowed.

Lambs weighing more than 10 kg can be dosed with the lamb-sized needle capsule. This will protect them from the growth-depressing effects of copper deficiency.

Cobalt

Cobalt is required in the rumen to facilitate the manufacture of vitamin B_{12} by the microflora. There is no effective storage of cobalt in the body, and ideally a constant supply should be provided.

In cobalt-deficient areas, dietary cobalt can be supplemented by: monthly oral dosing with a cobalt solution or trace element-enriched anthelmintic, treatment of pasture with a cobalt solution or use of the glass bolus which also contains copper and selenium.

Figure 6.4   Capsules containing copper oxide needles (one opened)

In past years the so-called cobalt 'bullet', a cylinder of cobalt alloy, given orally, proved an efficient way of preventing cobalt deficiency. The bullet lodged in the reticulo-rumen and slowly released cobalt. However, at the time of writing it appears that this useful product is to be withdrawn.

Oral dosing

Cobalt can be given orally on a monthly basis either alone or in a supplemented anthelmintic. This method has obvious limitations. On extensive grazings it will be undesirable to gather the lambs monthly.

Anthelmintic should be used only when required and a dose of anthelmintic should never be given just to satisfy a lamb's cobalt or selenium requirements.

On some lowland farms, monthly dosing with cobalt solution may well be feasible, and will be the cheapest way of preventing deficiency.

Pasture treatment

Dietary cobalt in herbage can be increased by dressing the pasture with a cobalt solution. The effectiveness of this measure can vary considerably depending on local conditions. Take advice from your agricultural adviser. This measure is normally impractable in extensive hill grazings.

Iodine

Iodine deficiency (p. 58) is relatively uncommon, and no specific agents are available for its prevention. Bought-in concentrate feeds and home mixes which incorporate a vitamin/mineral supplement should be adequate sources.

If additional supplementation is required, iodine salts must be given orally. Consult your veterinary surgeon about this.

Selenium

Selenium deficiency in ewes, with or without vitamin E deficiency, leads to stiff lamb disease (p. 87). This can be easily avoided by supplementing the ewes.

Bought-in concentrate feeds and home-mixed feeds which incorporate a vitamin/mineral supplement should be adequate sources, but often these feeds are fed too late in pregnancy to prevent problems, especially in the early lambing ewes.

In the past, the selenium 'bullet', a cylinder of selenium-containing alloy given by mouth, akin to the cobalt 'bullet', has been an effective way of preventing selenium deficiency. However, it seems likely that this product will become unavailable and alternative products must be used. Two long-acting preparations are currently available: selenium injections and the glass bolus which also contains copper and cobalt (see below). Selenium is available in injectable form either alone or with vitamin E (see below). To protect newborn lambs, the injection should be given to ewes after three months of pregnancy. In high-risk situations it may be advisable to inject newborn lambs soon after birth with a similar preparation.

Excess selenium is highly toxic to sheep. Fatalities have occurred when ewes have inadvertently been given excessive doses. Take care, a double dose does not have double benefit.

Combined preparations

Glass bolus

Some years ago the glass bolus incorporating cobalt, selenium and copper was introduced under the trade mark 'Cosecure'. The bolus was given by mouth, coming to rest in the reticulo-rumen. Here it was designed to slowly release adequate but safe quantities of the three essential trace elements. Unfortunately, the copper component did not perform adequately and the device was withdrawn.

However, the glass bolus, with the same trademark, has been reformulated and is now available again in forms to suit both ewes and lambs (Fig. 6.5).

The ewe bolus is best administered just before tupping and the lamb bolus can be given from two months of age. The device should provide adequate supplementation of the three trace elements for a whole season.

As has been noted, both copper and selenium are potentially toxic and no other trace element supplementation should be given, i.e. injections or copper oxide needles.

Vitamin E/selenium injection

When stiff lamb disease has been shown to be related to deficiencies of both vitamin E and selenium, this preparation is appropriate.

Ewes should be injected after three months of pregnancy.

In high-risk situations, a similar preparation can be given to newborn lambs. However, this must not be regarded as a substitute for supplementing the ewes.

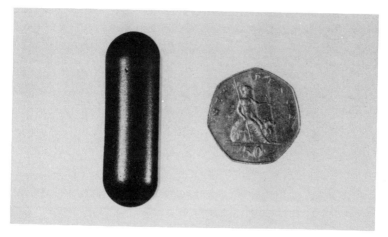

Figure 6.5   Cosecure glass bolus containing copper, selenium and cobalt

Combined vitamin injections

There are a multitude of combined vitamin injections on the market. The only ones generally of use to the sheep farmer are those containing vitamins A, D and E.

Where past experience has indicated that a deficiency of these vitamins is likely, and especially if the quality of preserved forage is poor, ewes should receive a supplementary injection at or just after mid-pregnancy.

Mineral blocks and powders

Free-access blocks and powders are commonly used, but are an unreliable way of mineral supplementation. Some ewes devour these products avidly whilst others never touch them.

The high iron content of these mixtures may have adverse effects on the absorption of other trace elements. Under no circumstances should blocks or powders designed for cattle be given to sheep. The copper content of these products may prove dangerous.

Vitamin/mineral drenches

A number of vitamin/mineral oral preparations are available. These supplements generally only give short-term relief of any deficiency. Repeated dosing is required to confer adequate long-term supplementation. These factors considerably reduce the practical usefulness of these products.

Chapter 7

Techniques for treating newborn lambs

Diagnosis of problems

The diagnosis of problems in newborn lambs depends on a knowledge of the problems likely to occur (Chapter 3), a careful examination of the lamb and the ewe (see below), and a consideration of the disease history of the whole flock. Some problems require no more than examination of the lamb, e.g. umbilical hernia, while others require a detailed investigation by your veterinary surgeon, and quite probably a veterinary investigation laboratory e.g. swayback. Whenever you are at all in doubt take professional advice – a stitch in time commonly saves more than nine.

Detection of sick lambs

The early detection of sickness in lambs contributes much to the success of treatment. It does, however, present a problem. Behaviour, appearance and response to a stimulus such as the presence of the shepherd vary considerably from lamb to lamb depending on age and type (single, twin etc.). There is no such animal as a 'normal' lamb with which to compare the potentially sick lamb. Experience helps, for this subject is just as much an art as a science. The only useful advice for the novice is 'if in doubt examine the lamb as described below'. In the end this will save time and lambs. Whenever you see a lamb curled up in the corner of a pen ask yourself: 'Is it sleeping off the effects of its last feed or is it sick?'

Examination of the sick lamb

The diagnosis of a problem in a sick newborn lamb depends on a careful examination. The temptation to jump at the apparently obvious symptom should be avoided – something of equal importance may be missed. Always follow the routine outlined below.

Before physically examining the lamb ask yourself the following questions:

1. How old is it? Many problems are age related (Table 7.1).
2. Was its birth assisted or protracted? It may have suffered severe hypoxia (high susceptibility to hypothermia), or it may have been injured (fractured ribs).
3. Is ewe thin or diseased? Lamb will have had little colostrum and may be starving.
4. Is the lamb very big? Birth problems likely.
5. Is the lamb very small? May be premature (susceptible to hypothermia).
6. Is the lamb weak AND unable to stand? A systemic or 'whole lamb' problem such as hypothermia.
7. Is the lamb strong BUT unable to stand? A problem affecting nerves or muscle such as swayback (Table 7.2).
8. Is the lamb unable to use both its hind legs? Swayback, spinal abscess or still lamb disease (Table 7.2).
9. Is the lamb lame on one leg? Fracture, joint ill or scad (Table 7.2).
10. Is breathing fast and/or heavy? Fractured ribs, pneumonia or prematurity (lungs poorly expanded).
11. Is the lamb's abdomen empty and tucked up? Starvation.
12. Is the lamb's abdomen swollen or blown? Watery mouth.
13. Has the lamb a poor birth coat? Prematurity or border disease.

Table 7.1 Problems in newborn lambs according to age at which they may first be seen

Birth
Atresia ani, Border disease, cleft palate, entropion, fractured ribs, jaw defects, joint defects, prematurity, umbilical hernia.

0–5 hours
Congenital swayback, daft lamb disease, fractures, hypothermia (exposure).

5–36 hours
Castration (incorrect), enteritis, hypothermia (starvation), inhalation pneumonia, stiff lamb disease, watery mouth.

36 hours
Eye infections, joint ill, lamb dysentery, liver necrosis, navel ill, scad, spinal abscess, tetanus.

The number of conditions to be considered increases as the lambs gets older, i.e. for a four hour-old lamb only conditions in the first two categories need be considered, but for a two day-old lamb all the conditions are possible.

Table 7.2 Conditions in which some abnormality of walking ability MAY
            BE the first symptom seen

Castration (incorrect)	Scad
Daft lamb disease	Spinal abscess
Joint defects	Stiff lamb disease
Joint ill	Swayback
Limb fracture	

Finally examine the lamb (Fig. 7.1).
1.   The anal area
     (a) Is the anus present? Atresia ani.
     (b) Is the lamb scouring? Enteritis, lamb dysentery.
     (c) Is the lamb's temperature:
          low? Hypothermia.
          high? Infection, hyperthermia if in a warmer.
     (d) Has the lamb been castrated correctly (rubber ring)?

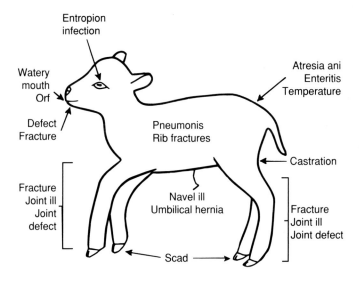

Figure 7.1   Examination of the sick newborn lamb

2. The trunk
    (a) Is the navel swollen? Navel ill.
    (b) Can you see intestines at the navel? Umbilical hernia.
    (c) Are there any skin wounds?

3. The head
    (a) Is there excessive saliva around the mouth? Watery mouth.
    (b) Is the lower jaw normal? Under- or overshot, fractures.
    (c) Is there any sign of orf - pustules or scabs in the lips?
    (d) Is the bottom eyelid(s) turned in? Entropion.
    (e) Is the eye(s) inflamed? Eye infection.

4. Legs
    (a) Are the joints fully extended? Joint defects.
    (b) Are any joints swollen? Joint ill.
    (c) Are there any other swellings? Fractures.
    (d) Is the cleft between the claws inflamed? Scad.

Taking the lamb's temperature

There are two types of thermometer available for recording a lamb's rectal temperature. They are basically used in the same way. Sit with the lamb on your lap, insert the thermometer/probe into the rectum to a depth of about 1½ inches (4 cm), allow time for the thermometer to warm up (normally about 30 seconds) and finally read the temperature.

Mercury clinical thermometer

This is an inexpensive glass thermometer commonly used in both veterinary and human medicine. It is easy to break and even easier to lose! It is made in such a way that it 'holds' the lamb's temperature after it has been removed from the rectum, unlike a normal room thermometer which goes up and down depending on the temperature of the environment. This means that the thermometer has to be re-set or zeroed before the next lamb's temperature is taken. This is easy to do by means of a flick of the wrist once you have got the knack – get your vet to show you. To read the thermometer you need good light – an inconvenience when working in the poorly-lit sheep house at night.

A special type of clinical thermometer, the sub-normal thermometer, is very useful when working with hypothermic lambs.

This instrument reads to a much lower temperature than does the normal instrument.

Electronic digital thermometer

This is convenient and easy to use. It consists of a rectal probe attached by a lead to a small plastic box which contains the electronics and the batteries (Fig. 7.2). The temperature reading is displayed on the front of the box in degrees Centigrade. Unfortunately these thermometers are not robust and can be damaged by water. One set of batteries is unlikely to last a full lambing season. The more expensive models have a light emitting diode (LED) display which is easy to read in the dark. Recently a single unit electronic thermometer has been introduced (Fig. 7.3) which has a liquid crystal display (LCD). In poor light this can only be used with the aid of a torch but its low price makes it an attractive proposition.

Interpretation of a lamb's temperature

Temperature	Interpretation
More than 40°C (104°F)	Fever–infection. If in a warmer, overheating (hyperthermia)
39–40°C (102–104°F)	Normal
37–39°C (99–102°F)	Moderate hypothermia
Less than 37°C (99°F)	Severe hypothermia

Note that a low temperature (hypothermia) does not necessarily mean that infection is absent. An infection such as joint ill may lead to starvation – net result a low temperature. It should also be noted that infections in lambs aged less than 24 hours rarely produce a fever (high temperature).

Feeding the newborn lamb

In general a stomach tube should always be used when feeding newborn lambs. A bottle and teat (a normal baby bottle is ideal but make the hole in the teat a bit bigger) is suitable for feeding the strong orphan lamb but can be lethal when feeding the weaker newborn lamb. Milk can easily enter the trachea (windpipe) and lead to inhalation pneumonia (p. 56).

However, it should be noted that it is dangerous to feed semi-conscious or unconscious lambs (normally hypothermic) with a

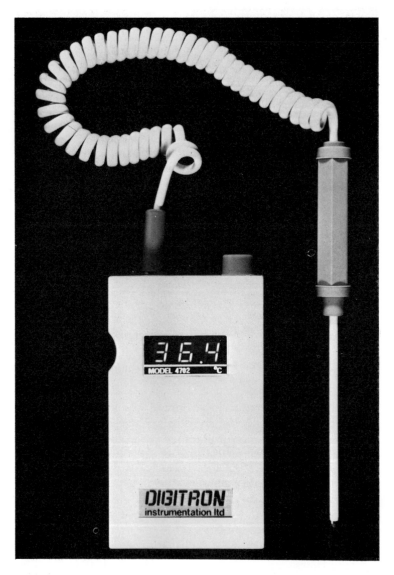

Figure 7.2   Electronic thermoeter (Model 4702, Digitron Instrumentation
Ltd). This model has a digital readout (LED)

stomach tube. In these lambs the tube can be easily passed into the
trachea and the lamb drowned. Even if the feed is correctly placed
absorption of nutrients is very slow and the food may even be

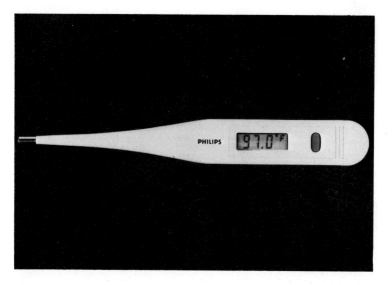

Figure 7.3   Electronic thermometer (Maximum thermometer, Type 5310, Philips). This model has a digital readout (LCD)

regurgitated and inhaled. If a lamb can lie in sternal recumbency (on its front), and hold up its head it is safe to feed it by stomach tube. If not, proceed as outlined in the hypothermia section in Chapter 3.

The equipment

Clean lamb stomach tubes and 60 ml syringes are required (Fig. 7.4). These should be rinsed after each lamb and sterilised at least once daily by immersion in a hypochlorite/detergent solution. This cleaning routine is most important and applies equally to all other feeding equipment. Dirty feeding equipment quickly becomes contaminated with bacteria and disease will be passed from lamb to lamb.

The feed

Ewe colostrum

This is the best food for the newborn lamb. When supplies are limited it should be restricted to the first one or two feeds.

If possible accumulate a store of ewe colostrum by milking ewes with a plentiful supply, e.g. ewes with single lambs. This can be stored in the deep freeze in small containers such as yoghurt pots or screw-topped plastic jars. Screw-topped plastic jars are ideal as they can be

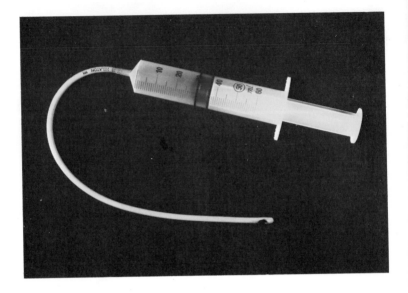

Figure 7.4   A rubber lamb stomach tube and a 60 ml syringe

immersed in a bucket of warm water for fast defrosting. Whatever the container, do not defrost frozen colostrum by boiling in a saucepan as this destroys the protective antibodies.

Cow colostrum

The best substitute for ewe colostrum is cow colostrum. This can be obtained from a dairy farmer, for whom it is a waste product, and stored in the deep freeze as already described. Cow colostrum does not, however, contain the same protective antibodies found in ewe colostrum. From the point of view of clostridial disease this problem can be overcome either by injecting the lamb with antiserum or by vaccinating the cow with clostridial vaccine before she calves. Consult your veterinary surgeon on both these possibilities before lambing. Very occasionally, problems occur in newborn lambs which have been fed cow colostrum. A severe anaemia (shortage of red blood cells) develops, characterised by weakness, shortage of breath and pale gums. If this occurs consult your veterinary surgeon who can save the lamb by transfusing blood from a ewe to the anaemic lamb. Don't feed the suspect cow colostrum to any more lambs.

Milk replacer

This is an acceptable food for the lamb aged more than 24 hours but should not be regarded as a substitute for colostrum.

Glucose/electrolyte solution

This solution is used for feeding lambs which have enteritis or watery mouth. In an emergency it can be used to feed any hungry lamb. Use one of the proprietary calf scour mixtures, but add powdered glucose to bring the concentration of glucose in the feeding solution to 10 per cent, i.e. 100 grams per litre. This glucose supplementation is not necessary in lambs aged three weeks or older.

Colostrum substitutes

Recently a number of colostrum 'substitutes' have come onto the market, some of cow and others of sheep origin. It would seem likely that those of sheep origin would be more efficacious, but to date we have no objective information upon which to base any judgement.

It should be noted that even a half-feed, i.e. 25 ml/kg, of colostrum obtained from another ewe or ewe colostrum preserved in the deep freeze is likely to be more effective than any substitute. This half-feed should be complemented by a half-feed of some other food such as cow colostrum or milk substitute, to ensure adequate energy intake.

Feeding routine

If a lamb is not sucking from a ewe it should be fed at least three times daily, e.g. 7 a.m., 3 p.m. and 11 p.m. at the following dose rates:

Large lamb – average single, about 5 kg; 200 ml each feed.
Medium lamb – average twin, about 3.5 kg; 150 ml each feed.
Small lamb – average triplet, about 2.5 kg; 100 ml each feed.

If it is practical, feed lambs more often. The quantity per feed should be reduced proportionally.

Using the stomach tube

1. Sit comfortably on a stool or straw bale with the lamb on your lap (Fig. 7.5).
2. Gently introduce a clean stomach tube (with no syringe attached) via the side of the mouth (Fig. 7.6). No force is required. In a large lamb all but 2–5 cm of the tube can be easily introduced. If the lamb shows signs of discomfort withdraw the tube and start again.
3. Once the tube is in place, observe the lamb. It should show no signs

Figure 7.5   A comfortable position for feeding a lamb by stomach
tube

Figure 7.6   The stomach tube in place

of distress and will probably chew the tube. This lack of discomfort proves that the tube is in the stomach.
4. Attach a syringe of colostrum to the end of the tube (Fig. 7.7). Empty the syringe slowly, taking about 20 seconds. Remove the empty syringe and attach a full one. Repeat this process until the full feed has been given.
5. Finally, remove both syringe and tube as a single unit and give the lamb freedom to move its head or cough if it so desires.
6. Wash and disinfect the tube and syringe.

Administration of drugs

Principles
Drugs are given for a variety of purposes by a variety of routes. Each time a drug is used, four requirements must be satisfied:
1. The drug must reach the site or sites in the body at which its action is required.

Figure 7.7   Giving a feed by stomach tube

2. The concentration of the drug at these sites must be high enough to achieve the desired result.
3. The drug must be given over a period long enough to achieve the desired effect.
4. Toxic (poisonous) side-effects, which all drugs have to a greater or lesser degree, must be avoided.

To achieve these aims the route of administration, frequency of administration and dose rate are specified for each drug and these must be adhered to.

The routes of administration commonly used in sheep are: topical, oral and by injection (parenteral).

The *topical route* is used when only a local action is required, such as on a wound or in the eye. Only preparations specifically intended for the eye should be used on this organ.

The *oral route* is generally employed when the drug is required to be active within the gut, e.g. antibiotic preparations for the treatment of enteritis. In human medicine, many drugs are taken orally as pills, tablets and capsules which are absorbed from the gut into the body system where their effects are required. In sheep medicine, when we want a systemic effect, rather than a local gut effect, we generally give drugs by injection. Drugs designed for oral use should never be given by any other route.

Drugs are given by *injection* when the effect is required within the body system, e.g. an antibiotic for the treatment of joint ill, or calcium solution for the treatment of hypocalcaemia in the ewe. The type of injection used depends on the drug in question and on the speed and duration of action required. The *intravenous route* gives the quickest effect since all the drug is immediately delivered throughout the body by the circulation. This route also gives the shortest duration of action, hence its use in anaesthesia where an immediate but short-term effect is desirable. Only certain drugs can be given by the intravenous route and this procedure should be used only by your veterinary surgeon.

A rapid effect is also achieved by injection into the peritoneal (abdominal) cavity – *intraperitoneal injection*. Practically the only indication for the use of this route in lambs is the injection of glucose solution to starving hypothermic lambs.

The *intramuscular route* (injection into a muscle) is commonly used in sheep. Absorption from this site is quite rapid – high concentrations of the drug will be found in the bloodstream within an hour or so of injection – while the injected drug retained within the muscle acts as a reservoir, continually releasing more drug. The effective duration of action of drugs given by this route varies from 12 hours to two or three days. Large volumes of drug cannot be given by the intramuscular route. Practically, the maximum volume for a lamb is 2 ml and for a ewe 10 ml.

An injection under the skin *(subcutaneous injection)* is employed when either a comparatively slow release of the drug is required or when the volume involved is too great for administration by the intramuscular route. This route is used for the administration of vaccines and antisera, and for giving calcium and magnesium solutions.

Topical application (eye)

Great care must be taken when applying eye ointments. The eye should not be touched by the fingers or by the ointment tube. The lamb must be firmly held (Fig. 7.8). Open the eye by drawing the eyelids apart with the fingers. Squirt ointment into the open eye from a distance of about half an inch and allow the eyelids to close.

Figure 7.8   Administration of eye ointment. Note: the tube is not touching the eye

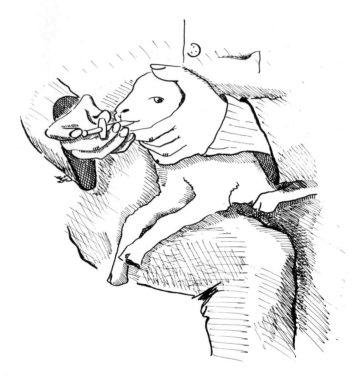

Figure 7.9   Administration of medicine by mouth using a syringe

Oral dosing

Drugs for oral dosing are often supplied in convenient dispensers. If not, a 2 or 5 ml plastic syringe should be used (Fig. 7.9). This should be gently placed over the lamb's tongue to ensure swallowing. Do not use an adult sheep drenching gun.

Oral preparations should never be given to lambs which are not fully conscious or are unable to swallow. In these lambs the drug, normally in liquid form, will either dribble out of the mouth or will enter the windpipe and cause inhalation pneumonia.

Injections

Equipment

Plastic disposable syringes and disposable needles are generally used nowadays. These disposable items are neither designed nor intended for repeated use and ideally should be used for one injection only.

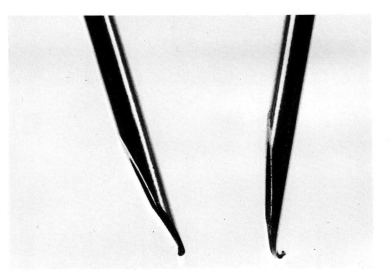

Figure 7.10 The points of two hypodermic needles which have been subjected to repeated use

When a syringe is used for the injection of antibiotic it may be used for one day, but after this it should either be discarded, or cleaned and sterilised by boiling before re-use. Disposable needles used for the injection of antibiotics should be discarded after about six injections and always at the end of the day. These needles quickly become blunt and barbed (Fig. 7.10), and further use will cause unnecessary pain and permanent damage to tissues.

When injecting non-antibiotic solutions such as glucose it is absolutely essential to use a new needle each time and either a new syringe or one that has been sterilised since the last injection. Solutions such as glucose are ideal media for the growth of bacteria and the repeated use of dirty equipment will result in serious, if not fatal, infections.

Subcutaneous injection

In the lamb the 'scruff' of the neck is the easiest site to use. Use either a 2 or 5 ml syringe, and a 1 inch 19 gauge needle. Pinch and raise a fold of skin and insert the needle into the fold holding the syringe at an angle of about 30° to the lamb's body (Fig. 7.11). Inject the solution and withdraw the syringe.

Figure 7.11 Injection by the subcutaneous route

Intramuscular injection

In the lamb the best site to use is the front of the upper hind leg – the muscle known as quadriceps femoris (Fig. 7.12). Use a 2 ml syringe and a 1 inch 19 gauge needle. Pinch the muscle mass between the thumb and index finger and insert the needle along the length of the muscle, almost parallel to the leg. Inject the solution and withdraw. This injection site is initially more difficult to identify than the muscles in the back of the leg, but it is a much more reliable site for drug absorption and there is little risk of nerve damage.

Many drugs for which intramuscular injection is denoted can safely and effectively be given by the subcutaneous route. Subcutaneous injections are easier and less painful for the lamb. Take advice from your veterinary surgeon for some injectable compounds must only be given by the intramuscular route.

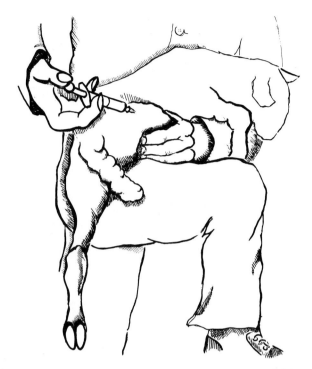

Figure 7.12   Injection by the intramuscular route. The injection is made into the front of the upper hind leg

Intraperitoneal injection

This route is only used for one type of injection – glucose solution for starving hypothermic lambs which have hypoglycaemia (low blood sugar level). The technique is described in full.

    Equipment required:
        Sterile 50 ml syringes
        New 1 inch disposable needles (19 gauge)
        20 or 40 per cent glucose (dextrose) solution (500 ml bottles)
        Electric kettle
        Antiseptic foot rot spray
    The dose to be given to a lamb depends on its size:
        Large lamb – average single, about 5 kg; 50 ml 20 per cent solution
        Medium lamb – average twin, about 3.5 kg; 35 ml 20 per cent solution

Figure 7.13 Injection of glucose solution by the intraperitoneal route to a starving, unconscious, hypothermic lamb

Small lamb – average triplet, about 2.5 kg; 25 ml 20 per cent solution

The solution for injection should be prepared immediately before use.

If using 20 per cent glucose solution, simply withdraw the required dose and warm to blood heat under a hot tap.

If using 40 per cent glucose solution, withdraw one-half of the required dose from the bottle and dilute this with an equal volume of

Figure 7.14   A lamb warmer constructed from bales

recently boiled water from a kettle. Shake the syringe and ensure that the solution is at blood heat. If recently boiled water is used this should result automatically.

To perform the injection:

1. Hold the lamb by the front legs as shown in Fig. 7.13.
2. Prepare the injection site (half inch to the side and 1 inch behind navel) by spraying with foot rot spray.
3. Fully insert the needle (with syringe attached) at the injection site with the needle tip aimed towards the lamb's rump (at an angle of about 45°).
4. Empty syringe and carefully withdraw. (The lamb may urinate during this procedure – this is not because the injection has gone into the bladder.)
5. Dispose of needle and boil syringe before re-use.

These notes are for guidance only. You must obtain professional instruction in this technique before using it yourself. Your veterinary surgeon may advise a precautionary injection of long-acting antibiotic at the same time as the glucose injection. Never administer an intraperitoneal injection to any lamb suffering a disorder of the gut, e.g. enteritis or watery mouth.

Warming hypothermic lambs

If a lamb's temperature is less than 37°C (99°F) it needs to be actively warmed. Infrared lamps are not advised because the rate of warming cannot be controlled, there is a serious risk of skin burns, and overheating (hyperthermia) can easily occur. The ideal way to warm hypothermic lambs is in air at 35–37°C (95–99°F). This temperature should not be exceeded. This warm environment may be obtained by making a 'bale' warmer (Figs 7.14 and 7.15), or by using the Moredun Lamb Warming Box (Fig. 7.16). The bale warmer is cheap but bulky, has a very slight fire risk and poor temperature control. The more expensive Moredun Lamb Warming Box is compact, avoids all risk of fire and has automatic temperature control. A third alternative is a home-made wooden box based on the concept used in the bale warmer (Figs 7.17 and 7.18). The dimensions of such a box should not be less than 1.5 metres square and 1 metre high, otherwise there is a serious risk of overheating. Heating for bale or box warmers should be provided by means of a domestic fan-heater with 1 kW, 2 kW and 3 kW output settings. Whatever type of warmer is employed it should stand on a layer of paper sacks to provide insulation. An ordinary household thermometer should be placed near the lamb to be absolutely sure that the correct temperature is being maintained (see above).

The lamb must be dried before it is warmed. The easiest way to do this is with a towel. If a wet lamb is placed in warm air it may lose more heat than it gains, due to the evaporation of water from its coat. The net result is that the lamb gets colder.

To warm a lamb, after drying and any other appropriate treatment (Fig. 7.19), place it in the top chamber of the warmer. Check the lamb's temperature at half-hour intervals and when it exceeds 37°C (99°F) remove the lamb from the warmer. The lamb will soon raise its temperature to normal by means of its own body heat. Feed the lamb by stomach tube (p. 155). Most lambs can now be returned to their ewes, ideally in a small sheltered pen. Take care that the lamb is well fed and does not become hypothermic again. One practical tip: when dealing with twins or triplets remove all the lambs while the hypothermic one is being treated and then after treatment return them together. This avoids rejection problems.

A few lambs will still be too weak after warming to be returned to their ewes. They may be unable to stand and suck. These lambs should be treated as described in the next section. We have summarised all the procedures for the treatment of hypothermic lambs in Fig. 7.19. In the

Plastic cover (1000 gauge), 2.15 m × 2.75 m, with wooden straps to weight it down. Adjust cover to control temperature.

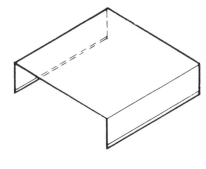

Top deck of six dry hay bales (straw may be used if hay not available).

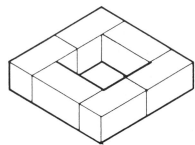

Lamb platform, 13 mm weld mesh, 1.5 m × 1.5 m.

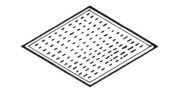

Bottom deck of six bales resting on layer of paper sacks for insulation. 3 kW fan heater with 1, 2 and 3 kW settings, placed between bales in steel safety tunnel (375 mm high × 450 mm deep × 600 mm wide). Adjust kW setting to control temperature. Leave thermostat at highest setting.

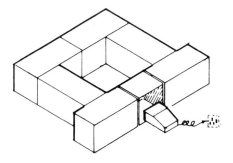

Figure 7.15 Plans for making a bale warmer. (From *Management at Lambing*, 1983)

Figure 7.16   A lamb warmer fitted with integral heater, fan and thermostat

United Kingdom the Agricultural Training Board run an excellent course on the treatment of hypothermia and this is recommended.

Care of the weak lamb

On occasion the shepherd has to care for a very weak lamb which cannot be left with its ewe. There are many causes of weakness including premature birth, high litter size (e.g. quads) and postnatal disease such as hypothermia or enteritis.

Figure 7.17 A lamb warmer of wooden construction based on the concept used in the bale warmer

The weak lamb has four basic requirements:
1. Treatment of any disease present.
2. Warmth.
3. Food.
4. Protection from infection.

To satisfy these requirements:
1. Treat any disease present.
2. House the lamb in an individual cardboard box or similar container under an infrared lamp (suspended about 4 feet above the lamb) (Fig 7.20).
3. Feed the lamb at least three times daily by stomach tube (p. 155). Colostrum should be given for the first day. Subsequently milk replacer can be used.

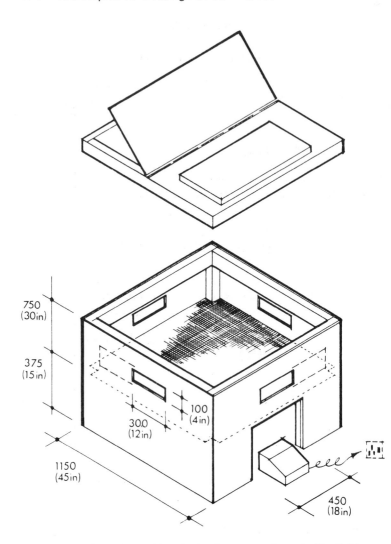

Figure 7.18 Plans for making the lamb warmer shown in Fig. 7.17. The lid has a transparent inspection window, the false floor is constructed using 13 mm (half-inch) weldmesh, and each of the four vents is fitted with a sliding, adjustable cover used to control temperature. (From *Management at Lambing*, 1983)

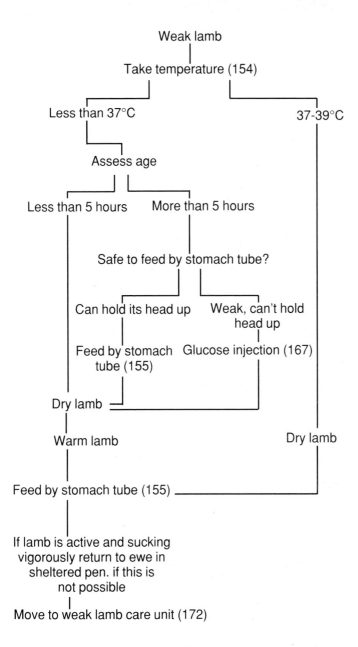

Figure 7.19   The treatment of hypothermic lambs (number references indicate the page where the relevant procedure is described)

Figure 7.20 Unit used to house very weak lambs. The infrared lamp is suspended about 4 feet above the lambs

4. Administer oral antibiotic twice daily – consult your veterinary surgeon about this.
5. Maintain the lamb in this system until it is stronger and free of disease. Then either foster to a ewe or rear artificially.

Artificial rearing

In most flocks there will be spare lambs which cannot be fostered (p. 181) and these will need to be reared artificially.

All too frequently these lambs are kept in a small, dark, damp shed and are fed at irregular intervals from a filthy washing-up liquid bottle fitted with a perished teat. Bacteria multiply in both the bedding and feeding equipment and the end result is a group of poor, sick lambs. This situation can be easily and profitably avoided by adhering to the principles contained in the following guidelines.

Management in the first three days

Ideally leave the 'spare' lamb with the ewe for the first 24 hours, and supplement the whole litter by stomach tube. Ensure that all the lambs receive plenty of colostrum. At about 24 hours of age lift the lamb – preferably the strongest – and transfer it to an individual cardboard box warmed by an infrared lamp (p. 172). Feed the lamb milk replacer by bottle three times daily. Allow the lamb 50 ml/kg each feed. Give oral antibiotic twice daily (consult your veterinary surgeon about this). At 72 hours of age transfer the lamb to the artificial rearing pen provided that it is strong, is sucking well and is showing no signs of disease. Never introduce a sick lamb – it will probably infect the others.

All lambs, irrespective of source or age, which are destined for artificial rearing, should undergo this 48-hour 'quarantine' period before being introduced to the artificial rearing pen.

Management in the rearing pen

Housing

Lambs should be reared in groups of up to twelve. Ideally, site the rearing pen in a covered yard. During the 'training' period (see below) the rearing pen should be restricted in size but once all the lambs are sucking well give them plenty of space – the more the better (Fig. 7.21). Move the feeding equipment daily to prevent a build-up of dung in one area. Provide straw bales arranged to form a cross. This ensures that the lambs can always find shelter from draughts. Move these bales twice weekly. Do not use infrared lamps, as they encourage the lambs to huddle in one spot on badly soiled bedding.

Feeding equipment

Use a lamb feeder (Fig. 7.22). The lamb will quickly become self-feeding, and the use of cold milk prevents short-term overfeeding.

Milk

Use a good quality ewe milk replacer. These are usually made up by mixing 200 grams of milk powder with water to produce one litre of milk. Check that the measure used to dispense the powder gives the correct amount. If too little powder is used, the lambs may starve; if too much, the lambs may become dehydrated and may also scour.

Training

1. Introduce lambs to the lamb feeder when they are expecting their next feed, i.e. hungry but NOT starving.
2. Ensure that there is milk in the teat.

Figure 7.21 A group of lambs in an artificial rearing unit. The lambs
have plenty of space

3. Gently hold the lamb on the teat and encourage it to suck.
   Squeezing the teat may help to give the lamb the right idea.
4. Repeat this procedure every few hours until you are sure that the
   lamb is sucking for itself.
5. If a lamb refuses to suck much milk, feed it by stomach tube – do
   not let it starve.
6. While training, keep the milk warm. This will encourage sucking.
7. Once training is complete feed the milk cold.

Milk requirements
   Individual lamb requirements vary considerably.
   4–5 days old: 500–750 ml/day
   6–12 days old: 750–1000 ml/day
   13 days old plus: 1.5–2 litre/day

Figure 7.22 A feeding bucket for artificial rearing, Dal. 58-10 teat
Lambar. (Picture by Dalton Supplies Ltd, Nettlebed)

Adjust the amount of milk so that there is always a little left over at the end of feeding time. This will ensure that the slower feeders get their full requirements. If the milk is restricted these weaker lambs will be in danger of starvation.

Hygiene
ALL feeding equipment must be rinsed, washed and sterilised each day. Use a hypochlorite/detergent solution.

Solid food

After about one week in the rearing pen introduce fresh hay and lamb pellets. Replace these feeds daily even if they have not been touched.

Water

Always provide clean, fresh water.

Weaning

If the cost of milk replacer were not a consideration, the determination of the best time for weaning would be a simple matter. One could safely suggest that lambs should not be weaned until they had reached a body weight of 15 kg. In the real world, where cost is a significant factor, it is inevitable that most lambs will be weaned at lower weights but care must be taken not to wean lambs too early, otherwise a serious check in growth will result. The following guidelines should help to prevent most problems:

1. Do not wean before 30 days of age.
2. Do not wean at a body weight of less than 10 kg.
3. Ensure that lambs are taking solid food before weaning.
4. Wean abruptly. Do not progressively reduce the milk allocation to a group of lambs. The big, strong lambs which are ready for weaning will continue to get milk, but the smaller lambs which are not ready for weaning will be weaned willy-nilly.
5. Assess readiness for weaning by relating present body weight to birth weight. A big single lamb, weighing 6 kg at birth, may need to be taken to 15 kg, whereas a small triplet, weighing only 2 kg at birth, could be safely weaned at 10 kg.

Health

The relatively close confinement of lambs in an artificial rearing system inevitably increases the risk of infectious diseases such as enteritis and eye infections. The incidence of these problems can be reduced by following these guidelines:

1. Give oral antibiotic for the first three days of life (take advice from your veterinary surgeon).
2. All lambs must undergo 48 hours of quarantine before introduction into the system.
3. Never introduce a sick lamb.
4. Watch the lambs closely and isolate and treat any sick lamb.
5. Clean and sterilise the feeding equipment daily.

6. Give the lambs plenty of space and fresh air while preventing draughts.
7. Avoid the use of infra-red lamps.

Urinary calculi can be a serious problem in ram lambs after weaning (see p.93). Calculi are small stones which collect in the bladder and eventually block the urethra, the tube connecting the bladder to the penis. The bladder eventually ruptures and the lamb dies. See page 94 for measures to avoid this problem.

Urinary calculi is a most painful condition for the lamb. If you suspect this problem – straining but little or no urine passed and tenderness of the lower abdomen – you should call your veterinary surgeon at once.

Fostering

In most flocks successful fostering is preferable to artificial rearing. It is, however, a far from foolproof technique and very high mortality rates are often recorded in fostered lambs. Guidelines are presented below which should help prevent some of these losses.

The lamb

Only strong, healthy lambs should be fostered. Inevitably lambs to be fostered will face problems and if attempts are made with either weak or sick lambs, failure can be expected.

The lamb is likely to come from one of two sources: a ewe which has too many lambs, e.g. triplets, or a poor ewe with twins; or out of the initial stages of an artificial rearing system (p. 176). If the lamb comes from a ewe with too many lambs, choose the strongest – not the weakest. If it comes from an artificial rearing system, take a lamb which has only been fed by stomach tube and has not become 'bottle orientated'. Whatever the source, the lamb must have received plenty of colostrum. During the fostering process ensure that the lamb never goes hungry. Hungry lambs soon become too weak to suck and are likely to be injured by the ewe.

The ewe

Only use a ewe as a foster mother if she is in good condition, has plenty of milk and is free of disease. In general it is better to avoid both very young and very old ewes.

Techniques

Four techniques are outlined: (1) rubbing-on at birth; (2) late rubbing-on; (3) lamb adopters; and (4) skinning. In our experience the rubbing-on techniques are the simplest and most effective.

Rubbing-on at birth

The success of this technique depends on speedy action after a ewe has had either a stillborn lamb or a single (check for the presence of another lamb by feeling the ewe's abdomen). The procedure is outlined below.

(a) Do not allow the ewe to rise to her feet after lambing.
(b) Rub the foster-lamb in the birth fluids, paying special attention to the anal region and the head.
(c) Tie together the lamb's front legs, so that it behaves like a newborn lamb and does not run around the pen.
(d) Place the foster-lamb plus the ewe's own lamb (if there is one) in front of the ewe and release her.
(e) Watch carefully from a distance but leave well alone.
(f) After an hour release the tied legs.
(g) For the next few days keep the ewe and lambs in a small pen. Check that the lambs are feeding and that the ewe is accepting them.

Late rubbing-on

On occasion it may not be possible to follow the procedure outlined above. Within about six hours of lambing a variation of the rubbing-on technique can be effective, especially if the placenta (afterbirth) has been retrieved.

(a) Place the ewe's own lamb and the foster-lamb in a CLEAN plastic dustbin.
(b) Throw the placenta on top of the lambs. If you have managed to save any birth fluids add these as well.
(c) Leave the lambs for an hour to 'mix'.
(d) Cut the placenta into halves and tie one half round each lamb's neck.
(e) Place the dustbin in the ewe's pen but do NOT release the lambs.
(f) After 30 minutes release the lambs.
(g) Watch carefully for signs of rejection and ensure that neither of the lambs goes hungry.

Lamb adopters

A variety of adopters can be purchased ready-made (Fig. 7.23), and plans for DIY models are available from agricultural advisors (Fig.

Figure 7.23 A steel fostering pen, MA1 Adopter Yoke. (Picture by
Modulamb Ltd, Coombefields)

7.24). All these devices comprise a small pen, measuring about 4 feet
square, fitted with a yoke in one side for restraining the ewe. The lambs
have the freedom of the pen and can suck when the ewe stands. Rails
are commonly fitted in the pen to enable the lambs to lie at the sides
without danger of being crushed.

The following procedure is followed.

(a) Place the ewe in the pen and secure her head.

(b) Put the lambs in the pen when they are due for their next feed, i.e.
hungry but not starving.

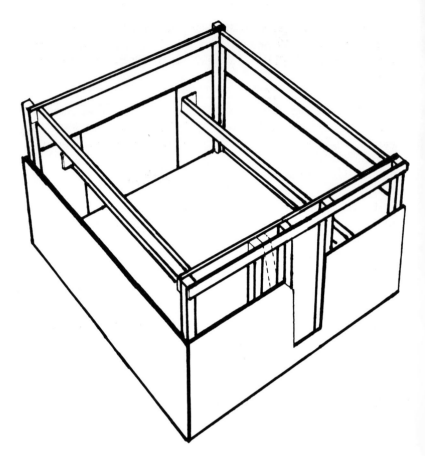

Figure 7.24 A wooden fostering pen. (From *Management at Lambing*, 1983)

(c) Leave for 48 hours. Check that the lambs are feeding. If they are not, encourage them to suck but if this fails, feed them by stomach tube. Do not let them starve.

(d) After 48 hours release the ewe from the yoke and watch carefully for signs of rejection, e.g. refusal to suckle, butting.

(e) If the ewe rejects the foster-lambs, give up. Further efforts are likely to be fruitless.

(f) If the lambs have been accepted release the ewe with her lambs into a small pen where you can watch them closely for the next few days.

This technique is most likely to be effective when the fostering process is commenced soon after lambing. It may be preferable to remove the ewe's own lamb (if she has one) and replace it with a pair of matched foster-lambs.

Skinning

Some shepherds swear by this technique, others consider it a waste of time. It should not be used when the ewe's own lamb has died from an infectious disease, such as enteritis. The procedure is as follows.

(a) Skin the dead lamb.
(b) Fit the skin to the foster-lamb.
(c) Put the ewe and foster-lamb in a small pen.
(d) Keep a close watch for signs of rejection and ensure that the lamb sucks. After 2 days remove the skin.
(e) If all is well after 3 days move the ewe and lamb to a small yard.
(f) If the ewe rejects the lamb, give up.

Summary

Irrespective of the fostering technique which you use, remember these four basic rules:

1. Do not use weak or sick lambs.
2. Do not use a ewe with insufficient milk.
3. Do not let the lamb(s) starve.
4. If at first you don't succeed, give up!

Castration

Three techniques are used: the rubber ring in the first week of life; the bloodless method at about 4–6 weeks of age; and the open or knife method. None of these techniques should be used without prior instruction. Incorrect or ill-timed use of the rubber-ring method of castration can cause problems (see faulty castration and watery mouth in Chapter 3). Notes on the correct use of this technique are given below.

Castration in the first 12 hours of life makes lambs more susceptible to watery mouth, probably by reducing colostrum intake. Leave castration until at least this age, and in the case of weak twins and all triplets until 24–48 hours, when you are sure that the lambs have had plenty of colostrum and are sucking well.

Why castrate?

At around five months of age the uncastrated ram lamb matures sexually and can become a considerable nuisance in the flock. If lambs are sent to slaughter before this age castration is not necessary – some would say it is undesirable as the entire lamb grows faster and produces a leaner carcass.

All sheep farmers should question whether castration is necessary in their lambs. If it isn't – don't do it.

Precautions

Whatever technique is used for castration ensure that the scrotum contains testicles, and testicles only. Any swelling of the scrotum suggests a hernia and possibly prolapsed intestines. Castration of such a lamb would likely prove fatal.

Castration and similar procedures should always be conducted in clean, dry conditions.

The law

1. No person aged under 17 years of age may castrate a lamb.
2. Only a veterinary surgeon may castrate a lamb older than three months.
3. Rubber rings may only be used in the first week of life. Use of the rubber ring after seven days of age is a criminal offence.

Rubber ring castration (Fig. 7.25)

1. Check that both testicles have descended into the scrotum and that the lamb does not have a scrotal hernia – intestines in the scrotum. A hernia can be felt as a soft mass within the scrotum. If in doubt mark the lamb and take professional advice. Castration would kill the lamb.
2. Place a clean, new ring well over the elastrator points, open the points by pressing the handles together, and with the points aimed towards the lamb, position the ring above the the testicles but below the teats (Fig. 7.25 and colour plates 11 and 12). If the ring is placed too high it may interfere with the urethra, the tube connecting the bladder to the penis. This prevents urination and will lead to the death of the lamb.
3. Release the pressure on the elastrator handles and check that both testicles are still within the scrotum.
4. Remove the elastrator.

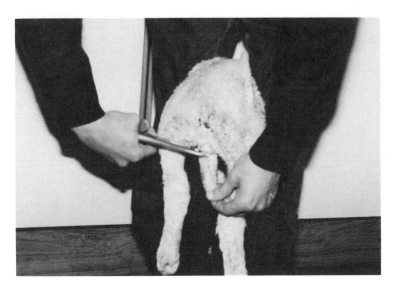

Figure 7.25 Castration using the rubber ring

5. Check again that there are two testicles below the ring and two teats above it.
6. If the ring has been wrongly placed it must be removed. Insert a blunt instrument, such as a teaspoon handle, under the ring and cut through the ring down onto the handle. This avoids any risk of cutting the skin.
7. Check all castrated lambs a few hours later ensure that they are mothered-up correctly and that none is showing signs of discomfort. If in doubt check the position of the ring again. The rubber ring is also used for docking the tail to avoid the danger of fly strike (Fig. 7.26). Short docking is unlawful and the unhealed wound that can result can increase the chance of fly strike. Sufficient tail must be left to cover the vulva in the ewe lamb and the anus in the ram lamb.

Future castration policy

Techniques for castration and docking are currently under review by the Farm Animals Welfare Council.

There can be no doubt that castration and docking by any means are painful for the lamb. An ideal solution to this problem would be to

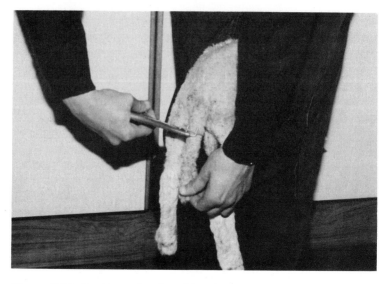

Figure 7.26 Docking using the rubber ring

employ some form of local anaesthesia, but there is some doubt as to whether a practical safe technique for use on-farm can be found.

We cannot anticipate the outcome of the Farm Animal Welfare Council deliberations but it does seem likely that the bloodless technique, suitable for lambs aged four weeks to three months, will be more widely used. Suitable equipment is shown in Fig. 7.27.

There is no doubt that this technique can be a safe, effective form of castration, but the potential for misuse and serious consequences for the lamb is very real. Before this technique is used full and proper training is required. A 'suck it and see' approach is to be condemned.

Navel dressing

The bacteria that cause joint ill, liver abscess, navel ill and spinal abscess commonly gain entry into the lamb through the wet navel cord (umbilicus) soon after birth, although the infections only become clinically evident a few days later. The incidence of these infections can be considerably reduced by keeping lambing pens clean and by 'dressing' navels as soon as possible after birth.

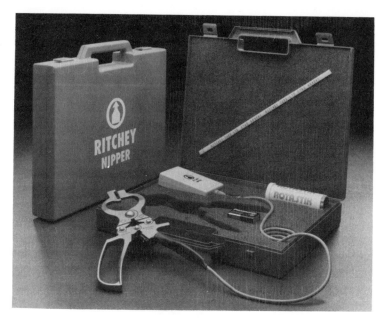

Figure 7.27   Bloodless castration equipment designed for use in lambs. (Picture by Ritchey Tagg Ltd, Masham)

The whole navel should be dipped in tincture of iodine. To ensure complete coverage, the jar containing the solution should be pressed against the lamb's belly and the lamb quickly upturned (Fig. 7.28). This solution contains two antiseptics – iodine and alcohol – and in addition the alcohol helps to dry the cord. Spraying the cord with an aerosol antiseptic or antibiotic preparation is unlikely to be as effective as this technique. It is sometimes suggested that the navel should be dressed again in the first day of life.

Giving an enema

An enema may be indicated in the treatment of constipation or watery mouth. This is easily performed using a 20 ml syringe and a cut-down stomach tube. Draw about 15 ml of warm soapy water (washing-up liquid in water) into the syringe and insert the tube about 5 cm (2 inches) into the rectum (Fig. 7.29). Inject the solution over about five seconds. In many cases faeces will be passed within 5–10 minutes.

Figure 7.28 Dressing the navel – the whole cord and surrounding skin are immersed

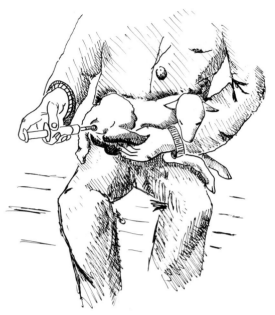

Figure 7.29   Giving an enema using a cut-down stomach tube and syringe

Chapter 8

Welfare at lambing

It is the prime responsibility of anyone, sheep farmer, shepherd or veterinary surgeon, to ensure the welfare of their stock at lambing, as at all other times.

There are two very good reasons for this. First the moral or ethical reason. Anyone who can knowingly cause distress or suffering to a ewe or lamb has no place on a sheep farm. Secondly, and this will be obvious to most in sheep farming, unhealthy sheep do not thrive. They certainly don't make profit. To the novice the detection of ill health and even pain in sheep can present a problem. The sheep but rarely makes an obvious response to pain or ill health. It adopts the 'dumb' approach. There may be good biological reasons for this, but for the shepherd or the sheep farmer it presents problems.

Changes in behaviour are often the first sign of trouble. To detect one oddly behaving ewe in a flock of 600 sheep requires keen unhurried observation and patience. This shepherding skill is only learnt in the field, not in the lecture room.

Early detection is the key to solving many problems, not only for the individual animal involved. A problem in one ewe (e.g. blowfly strike), may foretell a potential problem for the other 599 ewes. Swift preventative action may be called for.

Welfare codes

The Ministry of Agriculture (and corresponding authorities in Scotland and Wales) publish *Codes of Recommendations for the Welfare of Livestock*. These codes are not law as such but they occupy a similar place in legal practice to that occupied by the *Highway Code*. Failure to observe the welfare codes could be used in evidence as tending to establish the guilt of anyone accused of causing suffering to animals. By law, all stockmen must have access to these codes.

Below we have reproduced sections from the sheep code of special relevance to lambing.

Health

7   Sheep should be regularly inspected for signs of injury, fly strike, illness or distress. Frequent inspection is required in intensive systems and in other systems during lambing, and in the period before and after clipping and dipping.

8   Any injured, ailing or distressed sheep should be treated without delay and veterinary advice sought when necessary. Provision should be made for the segregation and care of seriously sick and injured animals. When a sheep has to be destroyed on the farm, this should be done humanely, and, where possible, by a person who is familiar with both the technique and the equipment used for slaughtering sheep.

9   Stockmen should be experienced and competent in the prevention and treatment of foot rot, the techniques of lambing, injection and oral dosing, tail docking and castration of lambs.

10   The health of flocks can best be safeguarded by the use of proper vaccination, foot care and dosing programmes based on veterinary advice.

11   Special care should be taken to ensure that all equipment used in dosing, vaccination and treatment is maintained to a satisfactory standard. Equipment used for any injection technique should be frequently cleansed and sterilised to avoid infections at the site of injection. Disposable needles should be used whenever possible. Dosing gun nozzles should be of a suitable size for the age of the sheep.

12   It is essential that all practical measures be taken to prevent or control external and internal parasitic infestations. Where infestations such as fly strike are likely to occur, sheep should be given routine treatment by regular dipping or other effective methods.

Management

Tail docking

21   The anal and vulval regions of sheep are sensitive areas and care must be taken to ensure that sufficient tail is retained to cover the vulva in the case of female sheep and the anus in the case of male sheep.

Castration

22   Castration must be carried out only in strict accordance with the law by a competent and trained operator.

Pregnancy and lambing

32   Scanning techniques can be useful to determine foetal numbers and the diet adjusted accordingly.

33   Heavily pregnant ewes should be handled with care to avoid distress and injury which may result in premature lambing.

34   Pregnant and nursing ewes should receive sufficient food to ensure the development of healthy lambs and to maintain the health and bodily condition of the ewe.

35   Stockmen should pay particular attention to cleanliness and hygiene. Every effort should be made to prevent the build up and spread of infection by ensuring the lambing pens are provided with adequate clean bedding and are regularly cleansed and disinfected. It is particularly important to ensure that dead lambs and afterbirth are removed and disposed of without delay, preferably by burial. There is a potential health risk to pregnant women from aborting sheep, those ewes at risk of abortion, dead lambs and afterbirths. Pregnant women should therefore stay away from sheep at lambing time.

36   It is vital that every newly-born lamb receives colostrum from its dam, or from another source, as soon as possible and in any case within 6 hours of birth. Adequate supplies of colostrum should always be stored for emergencies.

37   Stockmen should be trained in resuscitation techniques such as feeding by stomach tube. Some form of heating should be available to revive weakly lambs. Where lambing takes place out of doors some form of shelter or windbreak should be available.

38   The problems of mis-mothering, which occur particularly during transport or dipping, can be avoided by keeping group size to a minimum. Careful marking of lambs and mothers may also be beneficial.

Artificial Rearing

39   Artificial rearing can give rise to problems and requires close attention to detail and high standards of supervision and stockmanship to be successful. It is essential that the lambs should be allowed to suck the ewe for at least the first 12 hours of life (see paragraph 36).

40 All lambs should receive an adequate amount of suitable liquid food, such as ewe milk replacer, at regular intervals each day during their first three weeks of life. Where automatic feeding equipment is provided lambs should be trained in its use to ensure an adequate intake of food. From the end of the second week of life, lambs should also have access to palatable and nutritious solid food (which may be grass) and fresh clean water.

41 Troughs should be kept clean and any stale food removed. Equipment and utensils used for liquid feeding should be thoroughly cleansed at regular and frequent intervals and should be effectively sterilised.

42 A dry bed and adequate ventilation should be provided at all times. Where necessary, arrangements should be made to provide safe supplementary heating for very young lambs.

43 For at least the first 3 weeks of life, housed lambs should be kept in small groups to facilitate inspection and limit the spread of disease.

44 Where young lambs are being reared at pasture, care should be taken to ensure that they have adequate shelter.

Housing

Buildings and equipment

50 All floors should be designed, constructed and maintained so as to avoid discomfort, distress or injury to the sheep. Remedial action should be taken if any of these conditions occur. Solid floors should be well-drained and provided with some form of dry bedding. Newly-born and young lambs should not be put on slatted floors unless suitable bedding is provided.

51 Water bowls and troughs should be constructed and sited so as to avoid fouling and to minimise the risk of water freezing in cold weather. They should be designed and installed in a way that will ensure small lambs cannot get into them and drown. They should be kept thoroughly clean and should be checked at least once daily and more frequently in extreme conditions to ensure that they are in working order.

Lighting

53 Throughout the hours of daylight the level of indoor lighting, natural or artificial, should be such that all housed sheep can be seen clearly. In addition, adequate lighting for satisfactory inspection should be available at any time.

Space allowances

55   When sheep are fed in groups, there should be sufficient trough space or feeding points to avoid undue competition for food.

Hazards

68   Young lambs should be protected as far as possible from hazards such as open drains and predators.

References

1. Training courses which follow the Code recommendations are arranged for stockmen by the Agricultural Training Board, Agricultural Colleges and local education authorities. Proficiency testing in relevant subjects is carried out in England and Wales by the National Proficiency Tests Council, and in Scotland by the Scottish Association of Young Farmers' Clubs.

3. The Welfare of Livestock (Prohibited Operations) Regulations 1982 (SI 1982 No. 1884 as amended by SI 1987 No. 114) prohibit tooth grinding, freeze dagging and short-tail docking unless sufficient tail is retained to cover the vulva in the case of female sheep and the anus in the case of male sheep.

4. Under the Protection of Animal Acts 1911 to 1988 (in Scotland, the Protection of Animals (Scotland) Acts 1912 to 1988), it is an offence to tail dock or castrate lambs which have reached the age of 3 months without the use of an anaesthetic. Furthermore the use of a rubber ring or other device to restrict the flow of blood to the tail or scrotum is only permitted without an anaesthetic if the device is applied during the first week of life. Under the Veterinary Surgeons Act 1966, as amended, only a veterinary surgeon or veterinary practitioner may castrate a ram which has reached the age of 3 months or dehorn or disbud a sheep, except the trimming of the insensitive tip of an ingrowing horn which, if left untreated, could cause pain or distress.

Sections 9 and 37 and reference 1 emphasise the importance of training. Lack of training leads to needless suffering and lost profit.

All sheep farmers should note section 55. All too often great care is given to ration calculation and formulation, but too little attention to feeding arrangements. The result is underfed ewes, mostly carrying twins or triplets. A ewe heavily pregnant with triplets may be carrying 20 kg (44 lb) of extra weight. She simply can't fight at the trough for her food.

All should note **reference 4.** This is the law, not a recommendation. To do otherwise is a criminal offence.

Causing suffering to animals is often conceived as an active offence, doing this or doing that. In the sheep world it is normally a negative offence, not taking action when it is required. Not feeding sheep is the most commonly reported offence.

Problems in lambs

Generally, welfare problems in lambs reflect a lack of prompt attention. Eye problems such as entropion are sadly a common example. The problem is present at birth but goes unnoticed. Only when infection and copious tears are evident one or two days later is action taken. The lamb has had to endure two days of needless suffering.

Feet are a continual source of suffering for sheep. If God had blessed the sheep with only two legs instead of four, approximately 5 per cent of all sheep would be off their feet, never to rise again. But sheep have four feet and struggle on despite obvious pain and discomfort. All foot problems must be treated thoroughly and promptly. Tomorrow won't do.

In closed flocks it should be possible to eradicate foot rot, to the benefit of both sheep's feet and the shepherd's back!

Tetanus, lamb dysentery and the other clostridial diseases should be history, and not part of contemporary sheep farming. If these diseases occur either vaccination has been missed, or the lamb has had insufficient or no colostrum. Again the problem is omission.

Incorrect castration is a problem related to training, or rather the lack of it. Checking for correct placement of the rubber ring is very simple, two teats above and two testicles below (p. 185), providing the operator has been correctly instructed. It is the sheep farmer's responsibility to ensure that this instruction is given.

Every lamb problem in Chapter 3 is a welfare problem, but prompt action will reduce pain and suffering to a minimum. Problems as diverse as fractures and faecal spoiling must cause untold misery. Both respond gratifyingly to appropriate treatment.

Problems in ewes

Problems in ewes start before lambing with pregnancy toxaemia. Medical treatment is far from perfect but much can be done from a nursing viewpoint. These ewes must be regularly turned, their legs massaged and joints moved.

Problems during lambing itself mostly stem from unduly prolonged attempts to deliver the lambs. Note the 'ten minute' rule in Chapter 2 (p. 22). Extended haphazard attempts may be eventually successful, but the chance of ewe and lambs being alive in two days-time is low.

After lambing mastitis and metritis are the two major infectious problems. Keen observation and early detection are vital if treatment is to be successful. Affected ewes are clearly unwell and must be rescued from their plight at the earliest opportunity. Don't forget that 'one jag is rarely enough'.

Humane destruction

It will sometimes be evident that a problem in a lamb or ewe is hopeless, and that death will be the eventual outcome. In other cases, veterinary treatment may be theoretically possible, but totally uneconomic.

In either of these circumstances the animal must be quickly and humanely destroyed. To abandon the animal, mumbling 'I've done all I can', is totally unacceptable.

In the case of lambs an overdose of anaesthetic is the best method. Talk to your veterinary surgeon about this. The vet may feel able to let you have one dose for use in an emergency. Don't attempt an intravenous injection. The best route is that used for the intraperitoneal injection of glucose solution in starving hypothermic lambs (p. 167). Remember such drugs are dangerous and must be safely stored. The carcass must be deeply buried or burnt – it could prove lethal to any animal, e.g. your dog, which ate it.

For adult sheep firearms are indicated, either a pistol or a rifle. A captive bolt pistol is the weapon of choice but rarely available on sheep farms. Aim as shown in Fig. 8.1 and have another bullet available should it be required. Take great care to protect other people from ricochet. Beware of confined spaces and concrete floors.

Shepherd welfare

It is in the interest of the sheep, the sheep farmer and the shepherd that employers give shepherd welfare a high priority.

Figure 8.1   Line of fire for the humane destruction of an adult sheep
with a pistol or rifle

A tired shepherd makes mistakes, and misses the vital early signs of disease. This is detrimental to stock, and also to the farmer's pocket.

The employment of extra labour at lambing is an increasing and most welcome development. But if relatively inexperienced staff are employed (e.g. students), they must have adequate supervision. They do not have 'the eye'.

Lambing equipment checklist

Equipment

Baby feeding bottles and teats
Cardboard boxes (for housing weak lambs)
Electric kettle
Hypodermic needles (19 gauge, 1 inch)
Infrared lamps
Lamb feeder (artificial rearing)
Lamb warmer
Lambing snare
Lambing ropes
Needle and tape for treating prolapses
Plastic dustbin (fostering)
Rectal thermometer
Syringes (2, 5 and 50 ml)
Stomach tubes
Thermometer for lamb warmer

Sundries

Disinfectant: non-irritant, e.g. Savlon
Disinfectant: hypochlorite/detergent
Disinfectant: general purpose
Frozen colostrum (ewe/cow)
Lamb milk replacer
Lambing lubricant
Polythene bags
Soap
Stock marker

Drugs (after consultation with your veterinary surgeon)

Antibiotic for use in the eye
Antibiotic for injection
Antibiotic for oral use
Antiseptic cream, e.g. Savlon
Antiseptic foot rot spray
Calcium solution for injection
Glucose (dextrose) solution for injection
Glucose (powdered)
Glucose/electrolyte preparation (for lambs with enteritis or watery
    mouth)
Lamb dysentery antiserum
Liquid paraffin
Magnesium solution for injection
Pulpy kidney antiserum
Tetanus antiserum
Tincture of iodine (25 g iodine, 25 g potassium iodide and 25 ml freshly
    boiled and cooled distilled water, made up to 1000 ml with 90 per
    cent alcohol)

Further reading

W.B. Martin and I.D. Aitken (eds). *Diseases of Sheep*. 2nd Edition. Blackwell Scientific Publications, Oxford, 1991.

Meat and Livestock Commission. *Feeding the Ewe*. MLC, PO Box 44, Queensway House, Bletchley, Milton Keynes, MK2 2EF, 1983.

M.J. Clarkson and W.B. Faull. *Notes for the Sheep Clinician*. Liverpool University Press, Liverpool, 1985.

A.W. Speedy. *Sheep Production, Science into Practice*. Longman, London, 1980.

D.C. Henderson. *The Veterinary Book for Sheep Farmers*. Farming Press, Ipswich. 1990.

Glossary

Abomasum. The functional stomach in the newborn lamb. The fourth stomach in the adult sheep.

Abortion. The premature birth of weak or dead lambs (normally associated with disease).

Abscess. A localised collection of pus in any part of the body, e.g. liver abscess.

Agricultural Training Board. A government organisation in the United Kingdom which promotes in-service training in agriculture.

Antibiotic. A substance which either kills or arrests the multiplication of bacteria.

Antibodies. Substances produced within the body which counteract infection.

Antiserum. A solution for injection which contains a high concentration of antibodies against a particular infection.

Ataxia. Muscular incoordination. Inability to coordinate voluntary movement.

Atresia. Absence of a normal opening in the body, e.g. the anus.

Bacteria. Free-living microorganisms, some of which can cause disease.

Bloat. An excessive amount of gas in part of the digestive tract.

Bolus. A large pill.

Caesarean section. A surgical operation performed under either general or local anaesthesia by which the lamb is delivered through the abdominal wall.

Carbohydrate. A sugar or starch, e.g. glucose.

Caruncles. Button-like structures on the inner wall of the uterus through which oxygen and nutrients are passed to the foetus via the placenta.

Cerebrocortical necrosis (CCN). A nervous disease of ruminants caused by a deficiency of vitamin B_1, (thiamine).

Colostrum. The first milk produced by the ewe, very rich in antibodies.

Commensal bacteria. Infection with bacteria which normally does the host no harm.

Congenital abnormality. A defect present at birth.

Cotyledons. The parts of the placenta which make contact with the uterine caruncles and serve to transfer nutrients and oxygen to the foetus.

Cull. Remove a ewe which is unfit for future breeding from a flock.

Dag. Clip excess wool from the area around the vulva and anus.

Egg. The unfertilised female germ cell which develops in the ovary.

Elastrator. Device used for the application of rubber rings used for either castration or docking.

Embryo. The early developmental stages of the fertilised egg.

Enzootic. An animal disease which occurs commonly in a particular geographical area.

Ewe-lamb. A female sheep aged one year, a hogg.

Fat. A tissue found throughout the body which is an important source of energy.

Foetal stillbirth. Death of a lamb before the beginning of the birth process.

Foetus. The developing lamb in the uterus. Description normally used from about five weeks after conception.

Gimmer. A female sheep aged two years.

Hernia. The protrusion of an organ, often intestines, through the body wall, e.g. umbilical hernia.

Hogg. A female sheep aged one year, a ewe-lamb.

Hypoxia. A shortage of oxygen.

Immunity. The acquisition of resistance to infectious disease by either vaccination or previous infection.

Immunisation. The acquisition of immunity by means of vaccination.

Larva. An immature roundworm. Plural – *larvae*.

Litter size. The number of lambs carried by a ewe through one pregnancy.

Meconium. The faeces which collect in the bowel before birth, the foetal dung.

Microflora. The small organisms which populate the reticulo-rumen – bacteria and protozoa.

Necrosis. The death of part or parts of an organ.

Nephrosis. A non-inflammatory degeneration of the kidney.

Notifiable disease. A disease which, if suspected, must be reported to the authorities. Failure to do so would be a criminal offence.

Oesophagus. The flexible tube connecting the mouth to the stomach – the gullet.

Oocyst. The fertilised infective egg of a protozoan parasite, e.g. *Coccidia, Toxoplasma.*

Osteodystrophy. Defective bone formation.

Ovum. A fertilised egg.

Ovary. The female reproductive organ which produces eggs and also female sex hormones.

Oviduct. The tube which transports the egg from the ovary to the uterus. Fertilization takes place in this tube.

Ovulation. The shedding of the egg or eggs from the ovary.

Parturient stillbirth. Death of a lamb during the birth process.

Pessary. A large tablet for local use in the vagina or uterus.

Placenta. The foetal membranes and cotyledons, the afterbirth. The placenta carries oxygen and nutrients from the uterine wall to the foetus via the umbilical cord (navel).

Pneumonia. Infection of the lungs.

Prematurity. The birth of a lamb before the normal length of pregnancy is complete (less than 140 days after conception).

Prolapse. The displacement of an organ through a natural orifice, e.g. prolapse of the uterus through the vagina.

Proteins. Nutrients essential to growth and normal development.

Protozoa. A single-celled microscopic organism, e.g. *Coccidia, Toxoplasma.*

Reticulo-rumen. The first and second ruminant stomachs, rudimentary in the newborn lamb.

Roundworm. A parasitic worm of the class Nematoda. Includes gut worms and lungworms.

Rumen. The second and largest ruminant stomach, but rudimentary in newborn lambs.

Septicaemia. A disease of the whole body caused by the presence of bacteria and their poisonous products in the blood.

Stillbirth. The birth of a dead lamb.

Strike. Attack on animal's skin by maggots derived from fly eggs, e.g. blowfly strike.

Toxaemia. A condition in which the blood contains harmful products.

Trachea. The windpipe.

Vaccine. A preparation which induces immunity to a disease in an animal without causing the disease, often given by injection.

Virus. A microorganism, much smaller than a bacterium, which causes disease. A virus is only able to multiply within living tissue.

Vitamin. An organic nutrient essential in very small quantities for health and development.

Index